Western Fertilizer Handbook

Third Horticulture Edition

Western Fertilizer Handbook

Third Horticulture Edition

Western Plant Health Association

Edited by

Jerome Pier and Dave Barlow

WAVELAND PRESS, INC.

Long Grove, Illinois

For information about this book, contact:
Waveland Press, Inc.
4180 IL Route 83, Suite 101
Long Grove, IL 60047-9580
(847) 634-0081
info@waveland.com
www.waveland.com

Reissued 2018 by Waveland Press, Inc.

10-digit ISBN 1-4786-3846-X
13-digit ISBN 978-1-4786-3846-9

Printed in the United States of America

7 6 5 4 3 2 1

Dedication

THOMAS A. RUEHR

1943–2009
Soil Science and Plant Nutrition Educator

This edition of the *Western Fertilizer Handbook* is dedicated to Dr. Tom Ruehr. Dr. Ruehr left a lasting legacy to students and professionals through his dedication to teaching. He received degrees from Ohio State University, Iowa State University, and his doctorate from Colorado State University. He began teaching at California State Polytechnic State University in 1974, eventually teaching nearly every course offered by the Earth and Soil Science Department. Dr. Ruehr was passionate about teaching. He was frequently sought as a speaker by professional agronomic groups because of his ability to explain technical topics and make them relevant to real-life situations. His wide-ranging expertise allowed him to bridge complex subjects, bringing enthusiasm and insight to whatever he was discussing. He received multiple teaching awards recognizing his excellence. At the time of his death he was teaching part time. His untimely passing leaves a large void both at the university and in California agriculture.

Foreword

The *Western Fertilizer Handbook, Third Horticulture Edition,* is a distinct departure from both the two previous Horticulture editions and the Agricultural edition. The third edition staff spent time defining "horticulture" before serious editing began. The diverse nature of commercial plant production in the western United States blurs the line between the terms "agricultural" and "horticultural." Most people involved in the western agrichemical industry believe that when one speaks of horticulture, that one is referring to the turf and ornamental market. This idea guided the editors to focus the third edition on presenting a useful reference for the turf and ornamental industry.

There are many changes in the third edition. The chapter order has been reorganized. All new line art, photos, figures, and formatting give the third edition an updated look. All references to food and fiber crops have been deleted and replaced with turf and ornamental examples. The soilless media and soil amendments chapters have been reorganized. Information from two important handbooks, *Water, Media and Nutrition: A Grower's Guide* by D.W. Reed, and *Growing Media for Ornamental Plants and Turf* by Handreck and Black, is included throughout the third edition as the first reference is no longer in print and the second reference presents information in metric units. Finally, the writing style is designed to present information clearly to a lay audience with only basic scientific knowledge while still being useful to the advanced field practitioner.

Many things have not changed in the third edition. This edition of the *Western Fertilizer Handbook* is intended to provide a wealth of practical information for growing plants and using fertilizers sustainably. In the field, the handbook can be found within easy reach of a workstation, in a classroom, or the backseat of a pickup truck.

This edition would have never gone to print without the inspiration of Dr. Thomas Ruehr, to whom the handbook is dedicated and whose passing motivated me to keep working on the third edition, and Pam Emery who was a driving force from the start and brought us nearly to the end.

Jerome Pier, Ph.D.
Agronomist

Acknowledgments

Many industry and academic professionals contributed to this edition of the *Western Fertilizer Handbook*. Thanks to Al Ludwick and the Second Horticulture Edition Editorial Committee. Thanks to the Western Plant Health Association (WPHA), especially Renee Pinel, Rich Cornett, Pam Emery, and Corrie Pelc. Thanks to Jen Welsch at Bookmasters. Thanks to Gregory Frank and Box.net. Many members of the WPHA Soil Improvement Committee (SIC) either wrote or reviewed content for the Handbook. Special thanks to these SIC members for their contributions: Keith Backman, Sebastian Braum, Mike Buttress, Nat Delavalle, Tom Gereke, Jim Gregory, Dave Holden, Michael Larkin, James Lovelady, Eric McGee, Rob Mikkelsen, Bruce Roberts, Allen Romander, Tom Ruehr, and Steve Spangler. Thanks to these major contributors: Mike Atkins, Steve Cockerham, Richard East, Dave Goodrich, Gordon Harada, Darren Haver, Mitchell Johns, J.B. Jones, Don Merhaut, Megan Perkins, Stu Pettygrove, and Jack Wackerman. Thanks to J.R. Simplot Company and Crop Production Services for allowing Dave and I, respectively, to devote so much time to this important task. Thanks to Dave Barlow for his commitment, expertise, and for keeping me going. Finally, thanks to Mr. Pickle's, especially #13 on sourdough with no onions.

Table of Contents

Introduction

Western turf and ornamental horticulture may be described in many ways—dynamic, aggressive, modern, changing, and diverse. Its trademark is the production of high-quality plant products and aesthetically striking landscapes. The climatic zones created by the interaction of the cool Pacific Ocean and dramatic mountain ranges allow a very diverse array of plants to be grown in the West.

Turf and ornamental professionals are under increasing pressure to recommend and use sustainable practices. By improving one's knowledge of the growth and development of plants and the media, water, and fertilizer used to grow them, the turf and ornamental industry can continue to produce the stunning landscapes the world associates with the Western U.S.

The *Western Fertilizer Handbook, Third Horticulture Edition,* is intended as a "first responder" reference for the turf and ornamental industry when dealing with fertilizer issues. Although it could be read cover-to-cover, it is more often used in a nonlinear fashion, entered from the index or table of contents.

The handbook is designed to provide basic, practical information on plant growth and development, plant nutrients, growing media, and water quality in the first five chapters. The middle section focuses on fertilizers, both synthetic and organic, amendments, and methods of fertilizer application. After an introduction to hydroponic techniques, the handbook ends with diagnostic techniques and nutrient management guidelines. The appendices gather useful tables and techniques for managing and working with fertilizers. Each chapter ends with suggestions for supplementary reading that allow one to explore a topic more deeply.

It is hoped that the collection of knowledge found in the *Western Fertilizer Handbook* will provide a useful set of tools for all those who work with and appreciate western horticulture.

The Western Plant Health Association

The Western Plant Health Association was established in 1923 as the California Fertilizer Association. Its purpose was, and still is, to promote progress within and without the fertilizer industry in the interest of maintaining an efficient and profitable agricultural and horticultural community. Activities of the association include developing and disseminating new information to its members and end users; encouraging environmentally safe and efficient use of fertilizer; supporting production-oriented research programs; promoting awareness of the beneficial uses of fertilizer in schools, colleges, and universities; and maintaining open lines of communication between industry, academia, and state and federal legislative and regulatory agencies.

Many of the mentioned activities are performed by the Soil Improvement Committee, established in the mid-1920s. The Soil Improvement Committee produced the first edition of the *Western Fertilizer Handbook* in 1953. Many years and editions later, hundreds of thousands of copies of the handbook have been sold throughout the world.

CHAPTER 1

Principles of Plant Growth

Growth is one of the fundamental attributes of living organisms. It is more than a mere increase in size and weight. Growth represents a progressive and irreversible change in form involving the formation of new cells, their enlargement, and maturation into tissues and organs.

All plants must have a supply of light, heat, water, oxygen, carbon dioxide, and mineral elements to grow. The sun is the main source of energy. The atmosphere supplies carbon dioxide and oxygen. The soil supplies water and mineral elements as well as oxygen and carbon dioxide for plant roots. Growth stops, starts, or is modified as environmental conditions change. This chapter will discuss the basic requirements for satisfactory plant growth and will introduce some of the concepts involved in the mineral nutrition of plants.

THE PLANT CELL

All living plants are made up of cells. The *cell* is the basic structural and functional unit of an organism, a microscopic chemical factory that absorbs and secretes materials. Plant cells transform light energy into chemical energy in a process called *photosynthesis*, which takes place within chloroplasts. Respiration and energy

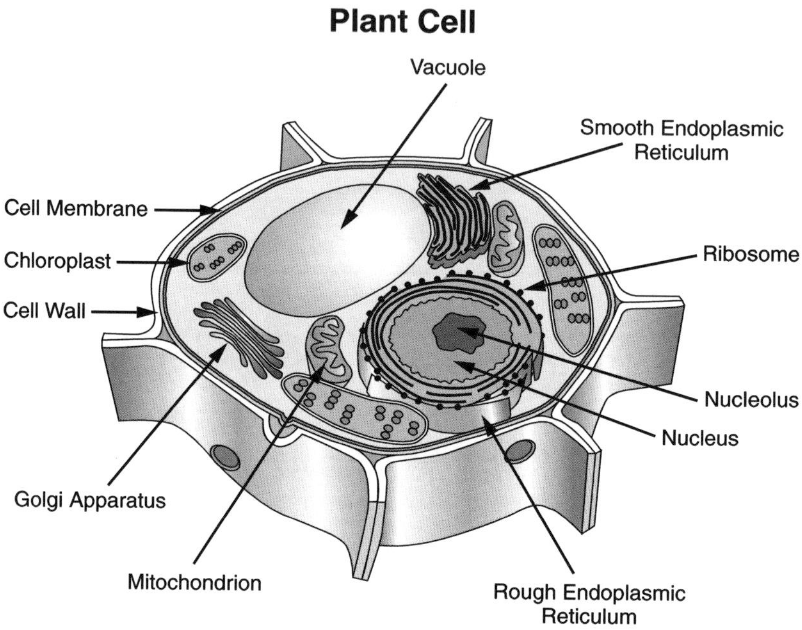

FIGURE 1-1. Plant cell and its major components.

transformations take place in the mitochondria of plant cells. Complex compounds are synthesized from the air, water, minerals, and simple sugars in the plant cells. The instructions for all plant growth and development are contained within the nucleus of each plant cell. A generalized plant cell is shown in Figure 1-1.

PLANT TISSUES

Groups of cells that function as a unit are referred to as *tissues.* Tissues may then be classified by function into one of four groups: meristematic, fundamental, protective, and vascular. *Meristematic tissues* are composed of the embryonic and undifferentiated cells capable of growth by division, which occurs at growing points. *Fundamental tissues* are made up of masses of cells that have little specialization in structure or function. Fundamental tissues act primarily as storage units. *Protective tissues* are the epidermal or outside surface of a plant. The function of *vascular tissues,* xylem and phloem, is to move water and compounds throughout the plant. These highly specialized tissues also add mechanical support because of their structure and location.

PLANT ORGANS

Groups of tissues form *organs*. Plant organs are generally separated into roots, leaves, stems, and reproductive structures. However, some plants may not have all of these organs or have other specialized organs.

Roots

The *root* is the plant organ that ordinarily grows downward into the soil, anchors the plant, and absorbs water, oxygen, and mineral nutrients. It may also serve as carbohydrate storage or as a reproductive organ. Roots vary greatly in form and size among species.

A close look at the root will show that it is made up of different functional areas. The primary region for absorbing water and mineral nutrients is the younger portion near the root tip. Root tips consist of the meristematic, elongation, and maturation zones (Figure 1-2). An even closer look at this region will reveal tiny root hairs radiating outward from the root surface several millimeters above the tip. Root hairs are about 0.01 millimeter in diameter and a few millimeters in length. Each hair is an outward extension of a portion of an epidermal cell. The outer, delicate walls of the root hair consist partly of gelatinous, pectic materials that enable the root hair to cling to soil particles and absorb water and nutrients. It is common to find many root hairs per square millimeter of epidermis in the root hair zone. Root hairs form in the elongation and maturation zones and later senesce and slough off as the root matures and lignifies.

Roots differ from shoots primarily in structure. Unlike shoots, roots do not normally bear leaves or buds and are not divided into nodes and internodes. Usually, roots differ from shoots in function and location but this is not always the case. Some plant roots develop buds that give rise to leafy shoots. Some plants have aerial stems that absorb water and nutrients.

The primary roles of older lignified roots are to conduct water and nutrients from the root tip to the shoot and transport

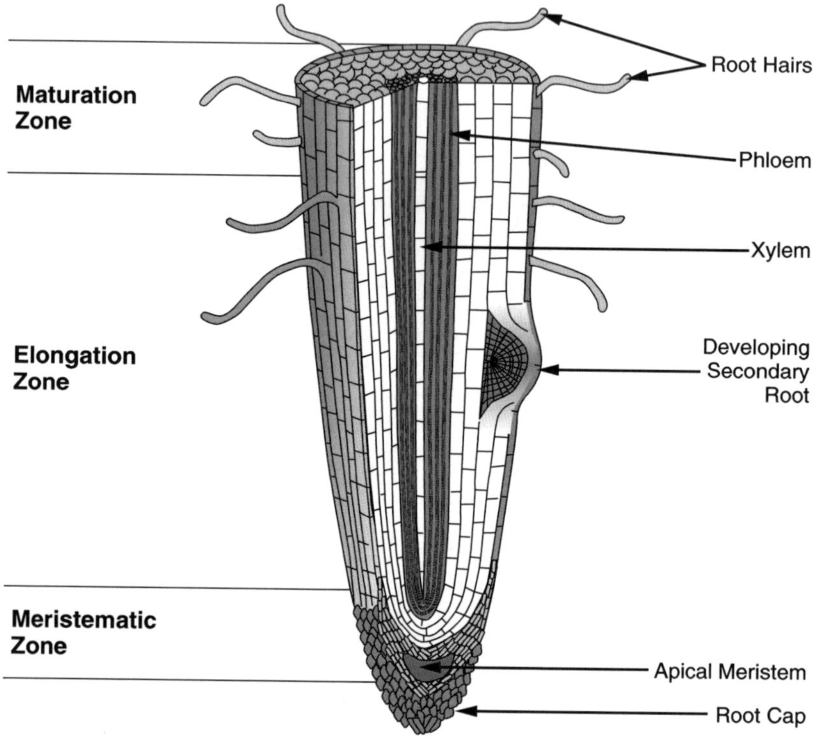

FIGURE 1-2. Microscopic view of a root tip.

synthesized compounds, such as amino acids and sugars, from the leaves back to the root tip. A cross section of a typical mature root is shown in Figure 1-3. The outer layer of cells, called the *epidermis,* provides protection. The root *cortex* provides support and storage. The *stele*, containing the conductive tissues, provides a means of transport for water and nutrients to and from the roots.

The soil zone immediately surrounding the plant root is known as the *rhizosphere*. This region has distinct differences from the soil farther from the root. Microbial activity is very high in the rhizosphere. The nutrient concentration in the rhizosphere may be lower as some immobile nutrients, such as phosphate, are depleted or higher as mobile nutrients, such as nitrate, accumulate in this zone. The water content in the rhizosphere may be lower due to transpiration.

Root systems are often grouped into two general categories: fibrous and taproot. *Fibrous* root systems consist of a group

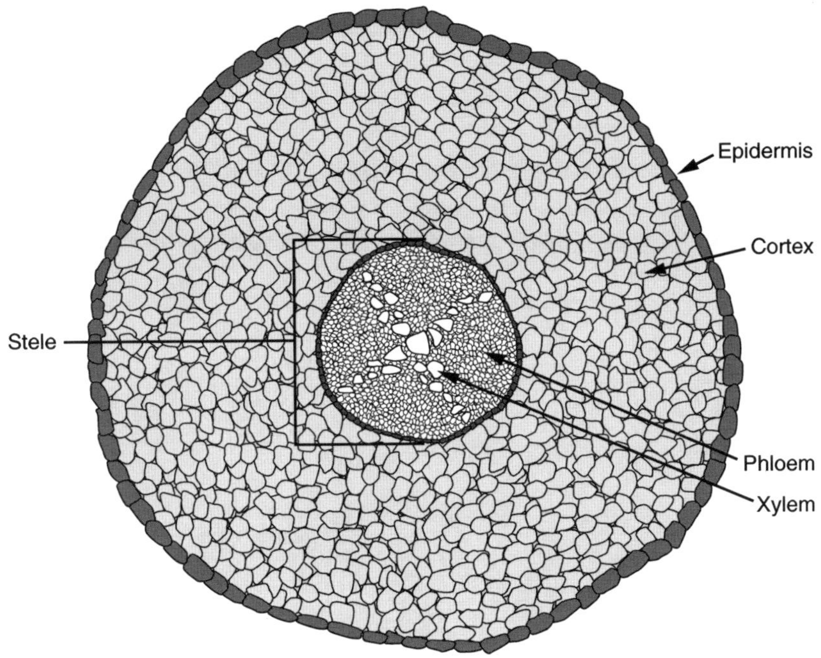

FIGURE 1-3. Cross-section of a typical root.

of numerous, long, slender roots of relatively equal width and length. Examples of this root system are berries, grasses, and many herbaceous and woody perennials. *Taproot* root systems consist of a large, downward-growing, primary root with smaller lateral roots developing from the primary. Examples of plants with taproots are dandelions, most broadleaf plants, and some trees such as conifers. Root systems not well defined as either fibrous or taproot are also common.

Root systems absorb and translocate water and nutrients, which require plant energy. This energy, which is derived from respiration, is needed to move ions against a concentration gradient. Most nutrient ions are present in higher concentrations within the root cells than concentrations found in the soil solution. Energy must be expended to move ions against this gradient.

Unlike most nutrients, water movement is primarily a passive, physical process. As water evaporates from leaves, it creates a difference in tension between the leaves and the roots. This tension "pulls" water up the plant. The pathway from initial entry of water through the active region of the root and through the cell walls resembles a long tube.

Shoots

Leaves and stems make up plant shoots. Each of these structures has separate but often overlapping functions. A generalized cross section of a leaf is shown in Figure 1-4. The leaf is the center of photosynthetic activity, the initial step in the carbohydrate manufacturing process.

The position of the leaf on the plant and the structure of the leaf facilitate trapping light energy by affording maximum exposure

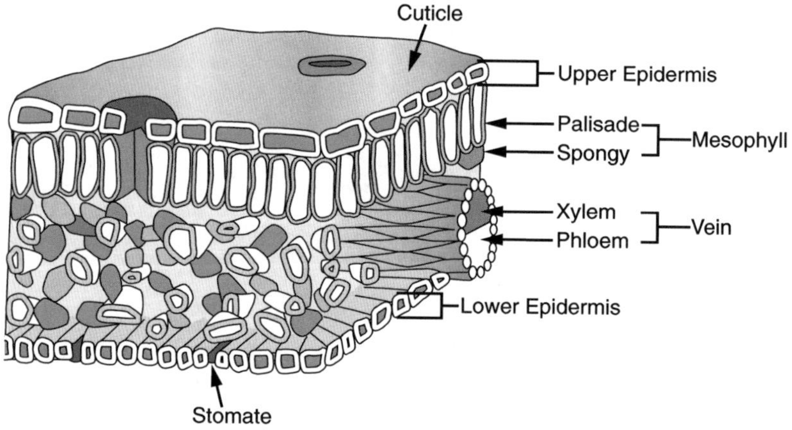

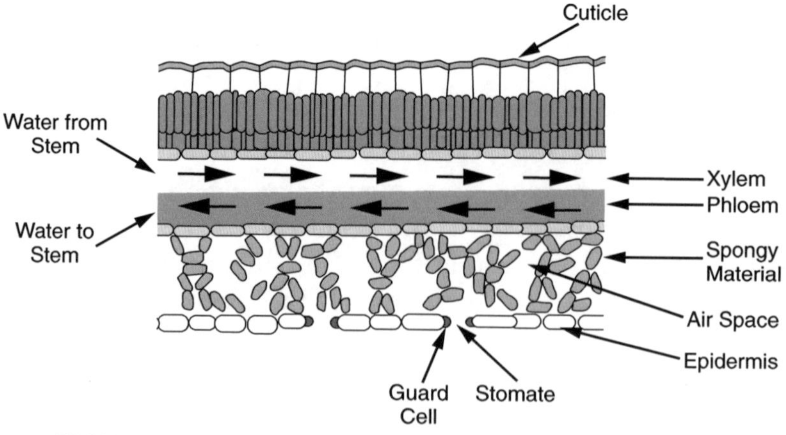

FIGURE 1-4. Cross-section of a leaf.

to the sun. Along the leaf surface are minute openings called *stomates,* or stomata, where exchange of carbon dioxide, water vapor, and oxygen occur. Underneath the epidermis is the *spongy layer,* composed of specialized cells containing chloroplasts. *Chloroplasts* are chlorophyll containing bodies that conduct photosynthesis.

The stems of some plants carry on photosynthesis if they contain chlorophyll. However, stems primarily contain the conductive tissues that carry water and nutrients to the leaves and transport sugars and starches to the roots and other organs. They may also serve as nutrient and carbohydrate storage for some plants. Stems support leaves, flowers, and fruit. Many stem modifications occur that are important to horticulture. Examples of modified stems are bulbs, stolons, rhizomes, corms, and spurs. Propagation of many species is accomplished through modified stems.

Reproductive Organs

The *reproductive organs* of the plant include flowers and the resulting fruits. The primary function of reproductive organs is to generate offspring through the production of viable seeds. Physiologically, most flowers utilize carbohydrates. In general, the larger the reproductive structure, the greater the potential sink it is for water, nutrients, and carbohydrates.

PHOTOSYNTHESIS

Photosynthesis is the process of converting carbon from atmospheric carbon dioxide and hydrogen from water into simple sugars. The energy required to drive the process is derived from light. Chlorophyll, the complex molecule found in the chloroplasts of green plants, is responsible for the reaction. The process occurs in all green tissues including stems, leaves, flowers, fruits, and sometimes the roots of some epiphytic plants. Since this process requires light, photosynthesis only occurs in the light. The generalized chemical equation for photosynthesis is:

$$\underset{\text{carbon dioxide}}{6CO_2} + \underset{\text{water}}{12H_2O} \xrightarrow[\text{chlorophyll}]{\text{light}} \underset{\text{glucose}}{C_6H_{12}O_6} + \underset{\text{oxygen}}{6O_2} + \underset{\text{water}}{6H_2O}$$

Carbohydrates produced by photosynthesis are transported to other plant parts, where they may be used immediately for cell maintenance and growth or stored after being converted to starch, sugars, fats, proteins, fiber, and other compounds.

RESPIRATION

Respiration is the process where energy, stored in carbohydrates produced by photosynthesis, is converted and stored in the energy rich molecule ATP (adenosine triphosphate). This energy is used for many different metabolic reactions, such as root and shoot growth, nutrient uptake, general plant metabolism, flowering, and seed development. Unlike photosynthesis, respiration occurs in all living tissues and occurs during both day and night. The fact that the roots must respire is a reason that oxygen is needed in the rhizosphere. Otherwise respiration and the resulting energy that is produced in the roots would not occur and the energy requiring processes of nutrient uptake and root growth would decrease or stop. The overall chemical equation for respiration is roughly the reverse of photosynthesis:

$$\underset{\text{glucose}}{C_6H_{12}O_6} + \underset{\text{oxygen}}{6O_2} \longrightarrow \underset{\text{energy}}{ATP} + \underset{\text{water}}{6H_2O} + \underset{\text{carbon dioxide}}{6CO_2}$$

TRANSPIRATION

The evaporation of water vapor from stomates on leaves, stems, and other aerial plant parts is called *transpiration*. Figure 1-5 illustrates the transpiration process. Most of the water absorbed by roots may be transpired to the atmosphere. Transpired water is vital to the life processes of the plant. The water that evaporates from leaves reduces plant temperature through evaporative cooling. Transpiration water loss also creates a tension within the conductive tissues that moves water, solutes, and other materials into the roots and the plant's vascular system.

The degree of transpiration that occurs is affected by the number and size of stomata (Figure 1-6) and whether the stomata are opened or closed. Waxy cuticles, thickened cell layers, decreased

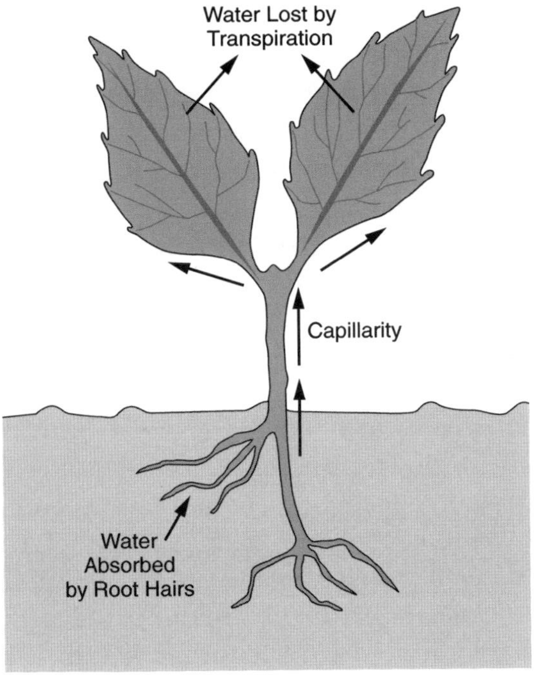

FIGURE 1-5. The transpiration process.

leaf surface area, reduced number of stomata, and hairy leaf surfaces are natural adaptations by plants that modify transpiration rates and allow for greater resistance to moisture stress. The opening and closing response of the stomata is related to changes within their guard cells and is affected by external and internal

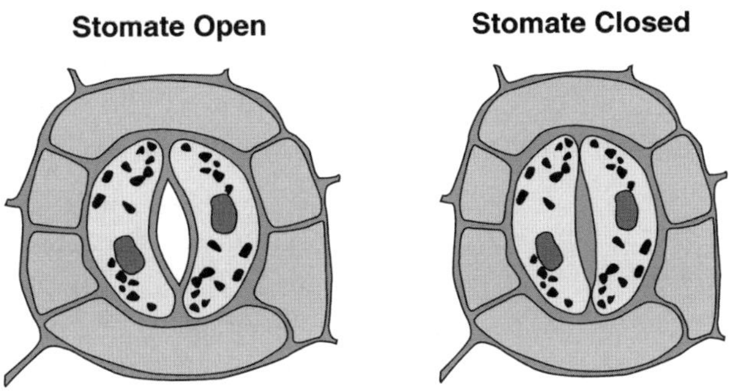

FIGURE 1-6. Diagram of open and closed stomates.

factors. External factors such as light, temperature, humidity, wind velocity, soil moisture, and salinity influence the opening and closing of stomata and thus, the rate of transpiration. Internal factors affecting stomata include the potassium content of the guard cells and the plant water status. Open stomata are more prone to water loss but they must be open to allow the gas exchange required for photosynthesis and plant growth. The rate of transpiration through stomata is greater during the day than at night.

NUTRIENT UPTAKE

Current information indicates that plants need seventeen elements to complete their life cycle. These essential elements are: carbon, hydrogen, oxygen, nitrogen, phosphorus, potassium, calcium, magnesium, sulfur, boron, chlorine, copper, iron, manganese, molybdenum, nickel, and zinc. Functions of nutrient elements in plants, their supply in the soil, and additions in fertilizers are discussed in Chapter 2.

Many more elements are found in plants, but their essentiality has not been established. Some of these, listed alphabetically, include: aluminum, arsenic, barium, bromine, cobalt, fluorine, iodine, lithium, selenium, silicon, sodium, strontium, titanium, and vanadium. This is not a complete list, since most known elements have been isolated at one time or another from plant materials.

With regard to nutrient uptake, there are two important physiological characteristics: active uptake and selective uptake. The primary reason that uptake requires energy is because nutrients are usually taken up against a concentration gradient. The concentration of nutrients inside the plant is often greater than their concentration in the soil solution. The energy required for nutrient uptake is derived from respiration. There are particular channels in root cell membranes that facilitate nutrient uptake. These channels are nutrient-specific, taking up and transporting specific elements across the cell membrane. This can occur because nutrient ions differ in size and charge. The special channels in roots can only accommodate nutrients of specific charges and atomic weights. However, some nutrients are chemically very similar, such as calcium and magnesium. If the soil solution concentration of magnesium is significantly higher relative to calcium, a magnesium induced calcium deficiency can result. Because of this phenomenon, it is important that both the concentration and ratio of the essential nutrients are maintained in the root zone.

GROWTH vs. TIME

The growth of a cell, an organ, or a whole plant does not proceed at a uniform rate. Growth starts slowly, gradually increases until a maximum rate is reached, and then slows. A characteristically S-shaped curve for an annual plant is obtained if total growth is plotted against time. This is illustrated in Figure 1-7. Fluctuations in temperature, moisture supply, or other environmental conditions may cause irregularities in the generalized trend, but if the whole period of growth is considered, the shape of the curve will remain essentially the same.

When nutrient uptake is plotted against time, the accumulation of many nutrients closely follows the growth curve. Note that nutrient uptake precedes growth, since nutrients must be present for growth to occur. A temporary shortage of nutrients will cause irregularities in the curve and a severe shortage will markedly slow growth.

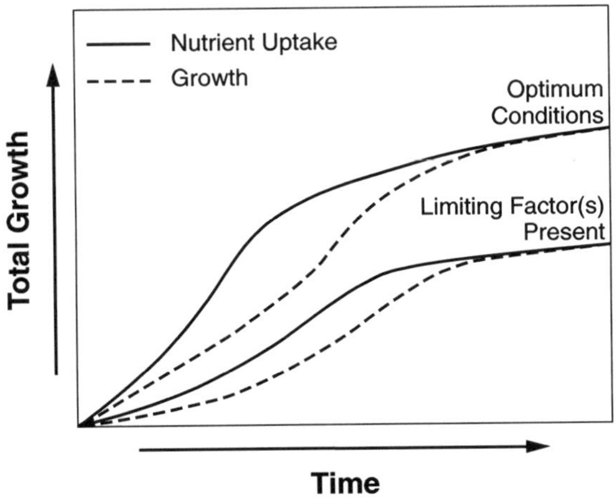

FIGURE 1-7. Characteristic nutrient uptake and growth curves for annual plants.

By necessity, this discussion of the many factors associated with plant growth has been simplified. Accordingly, much of the basic biology, chemistry, and related scientific disciplines have only been touched upon. Anyone desiring to become better acquainted with the subject of plant growth is referred to the suggested readings at the end of this chapter.

FACTORS AFFECTING GROWTH

How fast a plant grows and the shape or form it assumes is determined by internal and external factors. Heredity is the predominant internal factor and environment is the major external factor.

Genetic Factors

The tendency for an offspring to display the characteristics of its parents is known as *heredity.* For example, in a poinsettia plant, the length and strength of the stem, the shape and texture of the leaf, the number of bracts, and the resistance to cold, drought, and disease, have been shown to be inheritable traits. The male and female sexual cells that unite to produce the offspring contain these specific characteristics and the offspring is a product of that union. *Genes* are minute pieces of protoplasm located on the chromosomes of the cells. Genes provide the blueprint for all developing cells. Plant height, leaf and flower shape, and color, as well as all other characteristics, are determined by the unique information stored in genes.

Gregor Mendel is known as the father of plant genetics. He developed the theory of inheritance. Using the garden pea, he was able to single out a particular characteristic, such as flower color or smoothness of the seed, and demonstrate how it was inherited. Although the science of genetics is very complex and modern molecular techniques are widely utilized, the painstaking methods established by Mendel are still followed in genetic experimentation.

Light

The amount, quality, and duration of light available to plants plays an important part in their growth and development. Horticulturists may manipulate light conditions for desirable plant characteristics.

Most plants are able to reach maximum growth at less than full light intensity. The amount of available light is often modified by the density of plant canopy and shading.

Light also affects the direction of growth, a phenomenon known as *phototropism*. Bending toward light is termed positive phototropism. Bending away from light is known as negative phototropism. Shoots exhibit positive phototropism and roots exhibit negative phototropism.

A plant's response to the hours of daylight or darkness is called *photoperiodism*. Plants are classified based upon their reaction to day length as being short-day, long-day, or day-neutral plants. Short-day plants flower only under short-day conditions. In reality, the length of the uninterrupted dark period is the controlling factor in photoperiodic response, not the day length. However, in the natural world, long days necessarily have short nights, and vice versa. Thus, it is conventional to describe photoperiodism in terms of day length. If grown under long-day photoperiods, short-day plants will not flower and will grow vegetatively. Examples of short-day plants are poinsettias and chrysanthemums. Long-day plants initiate their flowering stage only when the length of day falls within certain limits, usually 12 hours or longer. Examples of such plants are carnations and hibiscus. Day-neutral plants, such as geraniums and roses, complete their reproductive cycle over a wide range of day lengths. Day length has also been reported to influence the formation of tubers and bulbs, the character and extent of branching, root growth, dormancy, and abscission or shedding of leaves and flowers. Photoperiodism plays a large part in the adaptability of plant varieties and species to other areas. A plant that grows well in one latitude may not thrive in other latitudes. Plant production systems can be set up to alter light received and induce flowering at any time of the year. Commercially grown poinsettias are forced to flower and produce the desired colorful display at Christmas by artificially altering the photoperiod in the nursery or greenhouse.

Gravity

Growth response to gravity is known as *geotropism*. Downward growth is the common response of roots to the pull of gravity.

Downward bending by roots is known as positive geotropism and upward bending by shoots is called negative geotropism.

Temperature

Growth response to temperature is known as *thermotropism*. Roots will generally grow in the direction of most favorable temperatures, thus exhibiting a positive thermotropism. Temperature has a marked effect upon plant growth and is one of the most important factors determining the distribution of plants over the earth's surface. The temperatures below and above which plant growth ceases are designated the minimum and the maximum temperature, respectively. An air temperature of 43°F is considered the lower threshold for most plant growth. The optimum temperature is the temperature at which plants grow best. Ideal plant growth is generally considered to occur between 65° and 75°F. Plant temperature thresholds may vary during the life of a plant and they also vary for different plant parts. Each plant species has a unique minimum, optimum, and maximum temperature.

Cold and heat tolerances have been established for horticultural crops to assist growers in selecting appropriate species for different growing regions. Temperature must be evaluated in terms of its direct effect on such basic processes as photosynthesis, respiration, water and nutrient absorption, and other chemical processes within a plant. Temperature also has an effect upon the biological and chemical transformations within the soil. Biological reaction rates double for every 18°F (10°C) increase in temperature. These transformations are intimately related to root activity, as illustrated by reduced uptake of phosphorus from a cool soil. Also, as soil temperatures warm, microbial activity is accelerated and nutrients are released from soil organic matter and plant residues.

Water

The importance of water to plant growth is readily apparent for many metabolic functions. Water is a requirement in photosynthesis, a part of protoplasm, and a vehicle for translocation of carbohydrates and nutrients. The availability and quality of water may influence the form, structure, and nature of plant growth. Chapter 5 discusses water and plant growth in great detail.

Growth response to moisture is known as *hydrotropism*. Roots will generally grow in the direction of a favorable moisture supply, demonstrating a positive hydrotropism. This characteristic is merely a response to the immediate environment. Roots and shoots do not seek out favorable temperatures or a nutrient supply, but merely grow better where these environmental conditions exist.

Atmosphere

Quantities of nitrogen, oxygen, and carbon dioxide do not vary in the atmosphere, except locally. Air normally contains 78 percent nitrogen, 21 percent oxygen, and 0.03 percent carbon dioxide. No higher plant can make direct use of elemental nitrogen without the action of certain microorganisms. All plants need oxygen from the atmosphere for respiration. All plants require carbon dioxide for photosynthesis. Increasing the carbon dioxide concentration in greenhouses has been shown to increase plant growth.

The atmosphere may also contain gases, particulate matter, and other substances that can have direct effects on plant growth. Some of these effects are positive while others are negative. For example, the aerial portions of some plants can absorb sulfur dioxide at low levels and much of the nutrient need for sulfur can be satisfied this way. Similarly, small amounts of atmospheric ammonia, a nitrogen source, can be absorbed through plant leaves. Specific damage to growing plants has been observed from air pollutants. Such compounds as ozone, excess sulfur dioxide, ethylene, fluorides, and others come from fuel combustion, organic solvents, and other sources. Damage from these pollutants has been noted on pine tree needles, flowers, and shrubs. Usually the injury shows up on leaves but sometimes the plants are stunted.

SUPPLEMENTARY READING

Biondo, R.J., and J.S. Lee. 2003. *Introduction to Plant and Soil Science and Technology*, 2nd ed. Interstate Publishers Inc., Danville, IL.

Capon, B., 1990. *Botany for Gardeners: An Introduction and Guide*, Timber Press. Portland, OR.

Farnham, D.E. 1985. *Water Quality: Its Effects on Ornamental Plants,* Rev. 1985. Division of Agriculture and Natural Resources University of California. Publication 2995.

Marschner, H. 1995. *Mineral Nutrition of Higher Plants*, 2nd ed. Academic Press, New York, NY.

Mauseth, J.D. 2009. *Botany: An Introduction to Plant Biology*, 4th ed. Jones and Bartlett Publishers, LLC., Sudbury, MA.

Mengel, K., and E.A. Kirkby. 2001. *Principles of Plant Nutrition,* 5th ed. Kluwer Academic Publishers, Dordrecht, Netherlands.

Raven, P.H., R.F. Evert, and S.E. Eichhorn. 1992. *Biology of Plants*, 5th ed. Worth Publishers, New York, NY.

Salisbury, F.B., and C.W. Ross. 1992. *Plant Physiology*, 4th ed. Wadsworth, Belmont, CA.

Taylor, H.M., W.R. Jordan, and T.R. Sinclair, ed. 1983. *Limitations to Efficient Water Use in Crop Production.* American Society of Agronomy, Madison, WI.

Wilkins, M.B., ed. 1984. *Advanced Plant Physiology*. John Wiley & Sons, Inc., New York, NY.

CHAPTER 2

Essential Plant Nutrients

There are more than 100 chemical elements known today. Only 17 elements have been shown to be essential to plants (Figure 2-1). These essential elements are required for normal growth, development, and reproduction and nothing can be substituted for them. Other elements may be found essential in the future. A few, such as cobalt and silicon, have already demonstrated an ability to stimulate plant growth under certain conditions. In 1989, nickel (Ni) became the most recent element declared essential to plants.

Three of the 17 essential elements—carbon, hydrogen, and oxygen—are obtained from air and water. The other 14 are normally absorbed from soil by plant roots. These 14 elements are divided into three groups: primary nutrients, secondary nutrients, and micronutrients. The primary and secondary nutrients are sometimes referred to as macronutrients. These groupings are based on historical patterns of observed nutrient responses and relative concentrations in plants.

CARBON, HYDROGEN, AND OXYGEN

Carbon (C), hydrogen (H), and oxygen (O) are supplied to plants from air and water. Carbon forms the skeleton for all organic molecules since plants are composed of 45 to 50 percent carbon by dry weight. Hence, carbon is a basic building block for plant life. Carbon is taken from the atmosphere by plants in the form of carbon dioxide.

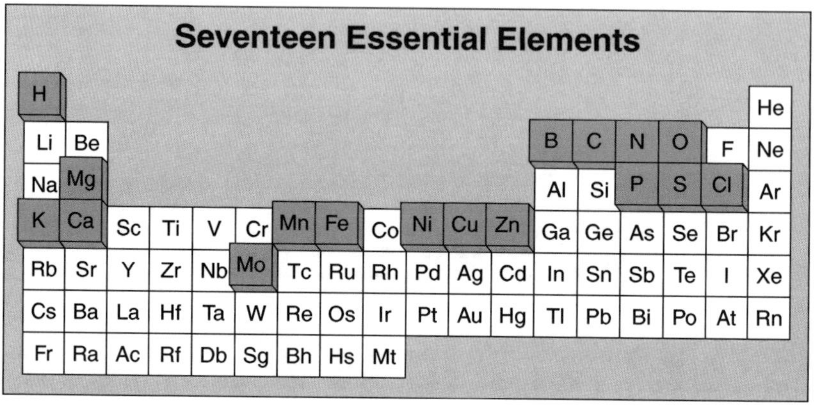

FIGURE 2-1. Periodic table of elements highlighting the 17 essential plant nutrients.

Through the process of photosynthesis, carbon is combined with hydrogen and oxygen to form carbohydrates. Further biochemical reactions, some with other essential elements, produce the numerous substances required for plant growth.

Respiration is the process from which plants derive energy by breaking down carbohydrates. Oxygen is required for plant respiration. Many compounds necessary for plant growth processes involve oxygen. Hydrogen, along with oxygen, forms water, which together constitutes a large proportion of the total weight of plants. Water is required for transport of minerals and organic compounds, and it also enters into many chemical reactions necessary for plant growth. Hydrogen is also a constituent of many other compounds necessary for plant growth. Since carbon, hydrogen, and oxygen are supplied to plants from air and water, it is not necessary to supplement the natural supply, unlike the other 14 essential elements.

Numerous tests have shown that there are times when supplemental carbon dioxide may be beneficial for plant growth. For example, greenhouses are typically closed during the winter to conserve heat. During daylight hours, the carbon dioxide concentration can drop considerably in the greenhouse reducing plant growth rates. For many greenhouse plants, carbon dioxide supplementation has been shown to be beneficial.

PRIMARY PLANT NUTRIENTS

Primary plant nutrients—nitrogen, phosphorus, and potassium—are usually the three most common deficient nutrients. Excluding carbon, hydrogen, and oxygen, plants take up more nitrogen and potassium than any other essential nutrient.

Nitrogen

Nitrogen in Plants

Nitrogen (N) is utilized by plants to synthesize amino acids, the building blocks of proteins. The protoplasm of all living cells contains protein. Nitrogen is also required by plants for other vital compounds, such as chlorophyll, nucleic acids, and enzymes. Of all the essential plant nutrients, nitrogen is most likely to be deficient in plants. Since both deficiency and excess of nitrogen can cause undesirable consequences, seek expert advice to determine the amount of nitrogen required.

Nitrogen is highly mobile within plants. When the soil nitrogen supply is inadequate, the young plant tissue becomes a strong sink, causing nitrogen to be withdrawn from the older leaves and transported through the phloem to the newer growth. As a result, deficiency symptoms first appear on older foliage. Potassium and magnesium are also mobile in the phloem and can have deficiency symptoms appearing first in the older growth.

Nitrogen is taken up by plants as nitrate (NO_3^-) or ammonium (NH_4^+) ions. Plants can utilize both of these forms of nitrogen in their growth processes. The growth of some plants is improved when the plants are supplied with both NO_3^- and NH_4^+ compared with either nitrogen form alone. However, high concentrations of NH_4^+ are generally avoided by many horticultural growers since it is thought to retard growth, decrease quality, or reduce the uptake of potassium, calcium, or magnesium. Research has indicated that the preference of plants for either NH_4^+ or NO_3^- is determined by the age and species of plant, the media properties, and other environmental factors.

In soil, nitrogen taken up by plants is frequently in the nitrate form. There are two basic reasons for this. First, nitrate is mobile in the soil and moves with soil water to plant roots, where uptake can occur. Ammonium, on the other hand, is adsorbed to the cation exchange sites of clay and organic matter and cannot readily move to the roots. Second, under proper conditions of temperature, aeration, moisture, and soil pH, soil organisms rapidly oxidize ammonium to nitrate. The process of converting ammonium to nitrate is called *nitrification*. In contrast, soilless media may have greater ammonium mobility and thereby allows plants to take up greater amounts of ammonium-nitrogen.

The choice of nitrogen form, nitrate, ammonium, or urea, is usually based on the potential for ammonium toxicity and the influence of the nitrogen form on substrate pH. Urea is considered similar to ammonium since it is rapidly hydrolyzed and converted to ammonium. When plants are provided with more nitrogen than can be utilized for growth, surplus nitrogen is stored within the tissue. When ammonium is accumulated in high concentrations in cells, it disrupts proper membrane function resulting in a condition known as ammonium toxicity. Nitrate can be stored in plant tissues in large concentrations without harmful effects. Ultimately, both ammonium and nitrate are converted into amino acids.

Symptoms of nitrogen deficiency in plants include:

- Yellow-green color of leaves (chlorosis)
- Slow growth; stunted plants
- Death (necrosis) of tips and margins of leaves, beginning with mature (older) leaves
- Needles yellowish, short, and close together in young conifers

The Nitrogen Cycle

The atmosphere contains approximately 78 percent nitrogen gas (N_2). It is estimated that over every 1,000 square feet of land, there are about 800 tons of nitrogen. In order for plants to utilize this nitrogen, it must be combined with hydrogen or oxygen. Breaking the strong chemical bond in nitrogen gas requires either the input of energy (to manufacture fertilizer) or specialized nitrogenase enzymes. This process is called *nitrogen fixation*.

Nitrogen may be fixed by specialized soil microorganisms. Some of these organisms live in nodules on roots of legumes and others are free living. The amount of nitrogen fixed by these organisms varies greatly, from a few pounds of nitrogen per acre per year in non-legume, natural soil ecosystems, to hundreds of pounds per acre per year where leguminous plants are grown. Lightning also fixes a small amount of nitrogen, which is carried into the soil by rainfall.

The fertilizer industry chemically fixes nearly 100 million tons of nitrogen each year into various nitrogen fertilizers. Fertilizer nitrogen is made by combining atmospheric nitrogen with a hydrogen source, usually natural gas, into ammonia. This reaction is referred to as the Haber-Bosch process. All other nitrogen fertilizers can be synthesized from ammonia.

The simplified *nitrogen cycle,* seen in Figure 2-2, shows the complexity of managing this nutrient. Since nitrogen is susceptible to many biological and chemical changes, proper management is essential to getting maximum efficiency.

Nitrogen may be lost to the atmosphere by denitrification of nitrate or by volatilization of ammonia. Leaching of nitrate, primarily from sandy soils, may move nitrogen below the root zone, where it cannot be utilized by plants. All are financial losses for the grower and may be environmental concerns to the community. Erosion of surface soil may also carry nitrogen from fields into streams and lakes or to the oceans. This continuous recycling of nitrogen is called the *nitrogen cycle*.

Nitrogen in Soil and Media

Most of the nitrogen in soil is a component of organic matter and is not immediately available for plant uptake. Only about 2 to 4 percent of this organically bound nitrogen becomes available to plants each year. Many soils of the western United States contain relatively small amounts of organic matter, so the amount of nitrogen made available for plant use each year may be relatively small compared with the nutrient demand of rapidly growing plants.

Many transformations involving nitrogen occur in soil, making it one of the most difficult nutrients to manage. Most of these transformations are the result of microbial activity. Nitrogen is made available for plant uptake from organic matter through two microbial processes. Proteins and N-rich compounds are broken down into amino acids through a process called *aminization*. Soil

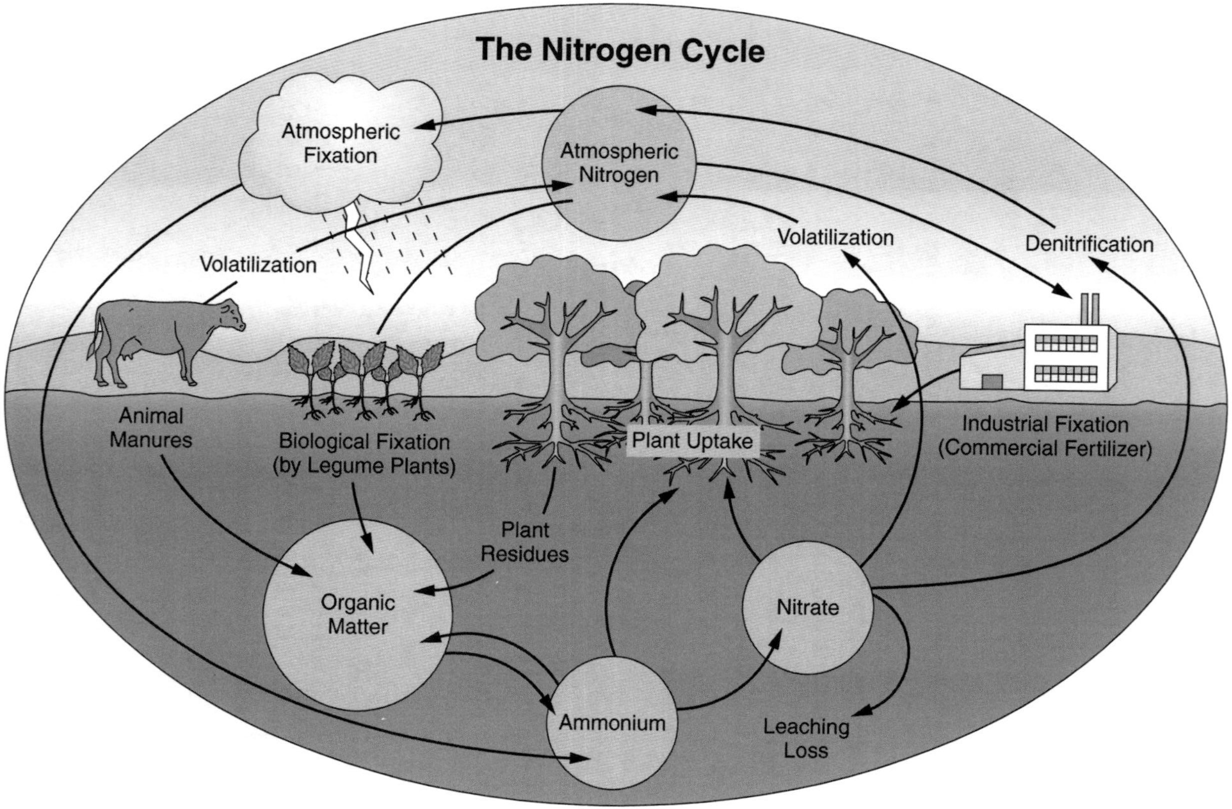

FIGURE 2-2. The nitrogen cycle.

organisms acquire energy from this process and utilize some of the amino nitrogen in their own cell growth. In the second process, called *ammonification,* amino compounds are converted into ammonia (NH_3) and ammonium (NH_4^+). The two processes, aminization and ammonification, are more commonly referred to as *mineralization.*

$$\begin{matrix}\text{Urea} & & & & \\ \text{Organic} & \rightarrow & \text{Ammonia} & \rightarrow & \text{Ammonium} \\ \text{Plant residues} & & & & \end{matrix}$$

After mineralization of organic matter or the application of ammonium fertilizer, ammonium is oxidized to nitrate by two distinct groups of bacteria in a two-step reaction. The ammonium may come directly from various N fertilizers or from mineralization of organic matter. The soil bacteria *Nitrosomonas* and *Nitrosococcus* convert ammonium to nitrite:

$$\underset{\text{ammonium}}{2NH_4^+} + \underset{\text{oxygen}}{3O_2} \rightarrow \underset{\text{nitrite}}{2NO_2^-} + \underset{\text{water}}{2H_2O} + \underset{\text{hydrogen ions}}{4H^+} + \text{energy}$$

In the second step, *Nitrobacter* then converts nitrite to nitrate:

$$\underset{\text{nitrite}}{2NO_2^-} + \underset{\text{oxygen}}{O_2} \rightarrow \underset{\text{nitrate}}{2NO_3^-} + \text{energy}$$

This two-step process in soil, called *nitrification,* occurs readily under conditions of adequate oxygen and moisture, warm temperature, and near neutral pH.

At 75°F, nitrification may be completed in one to two weeks (Figure 2-3); at 50°F, it may take 12 weeks or more; and at even cooler temperatures, it may require much longer. Since nitrate is mobile in soil and moves freely with water, considerable amounts of nitrate can be leached from the soil root zone without careful management. This can be a concern in fertilized turf, nurseries and horticultural operations. When using soilless media, nitrogen leaching may include the ammonium form as well as nitrate, since the cation exchange capacity is variable. Any addition of ammonium fertilizer to any media must take into consideration potential nitrogen leaching losses.

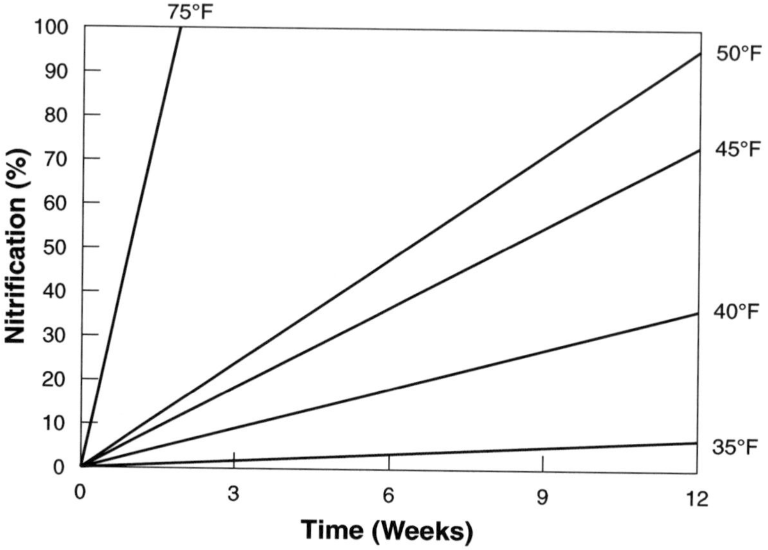

FIGURE 2-3. Generalized nitrification rates at various soil temperatures.

Nitrogen may be lost from the soil to the atmosphere by a process that converts nitrate to gaseous compounds of nitrogen. This process is called *denitrification*. Under anaerobic conditions (very low or no oxygen) caused by excessive moisture and/or compaction, certain bacteria are capable of utilizing oxygen from nitrate to meet their respiratory needs. In this process, various gases, such as nitric oxide (NO), nitrous oxide (N_2O), and nitrogen (N_2) are formed. As these gases move into the atmosphere, there is a loss of plant-available nitrogen from the soil. Nitrous oxide loss has also become an environmental concern since it is a contributor to global greenhouse gases.

The denitrification reaction can be generalized as follows:

$$\underset{\text{nitrate}}{NO_3^-} \rightarrow \underset{\text{nitrite}}{NO_2^-} \rightarrow \underset{\text{nitric oxide}}{NO} + \underset{\text{nitrous oxide}}{N_2O} \rightarrow \underset{\text{nitrogen}}{N_2}$$

Denitrification is not likely a major nitrogen loss for potted and greenhouse plants when proper drainage and soilless media are applied. However, denitrification can be a significant factor whenever plants are grown in soil, especially warm soils with poor drainage, and significant amounts of nitrate and soluble organic matter.

Nitrogen can also be lost to the atmosphere through ammonia *volatilization*. When ammonium-containing fertilizers, manures, and composts are applied to a neutral or alkaline pH soil, ammonium can be chemically converted to ammonia gas by the reaction:

$$\underset{\text{ammonium}}{NH_4^+} + \underset{\text{hydroxide}}{OH^-} \rightarrow \underset{\text{water}}{H_2O} + \underset{\text{ammonia gas}}{NH_3\uparrow}$$

Ammonia volatilization can be significant whenever ammonia-based fertilizers, including urea, or animal manures are surface applied. To prevent or minimize this loss, surface-applied ammonium fertilizers should be incorporated or irrigated into the soil soon after application. The soil should be wetted after incorporation if possible. Turf should be irrigated immediately after the broadcast application of ammonium fertilizers to move the fertilizer below the surface. Chemical additives, such as urease inhibitors, are available to delay the loss of ammonia from surface-applied urea.

Atmospheric ammonia is an air quality concern in many locations. When emitted to the air, ammonia reacts with oxides of nitrate and sulfate to form very fine particles that can cause respiratory problems for some people.

Phosphorus

Phosphorus (P) is present in all living cells. It is utilized by the plant to form nucleic acids (DNA and RNA) and many other vital compounds. Through energy-rich linkages (ATP and ADP), it is involved in the storage and transfer of chemical energy used for growth and reproduction. Phosphorous is highly mobile in plants.

Phosphorus stimulates seedling development, root formation, and flowering. It hastens maturity and promotes seed production. Phosphorus supplementation is required most by plants under certain conditions.

- During cold weather when phosphorus solubility is low
- With limited root growth due to factors such as cool soils, drought, toxic elements, or restricted rooting zones
- With rapid vegetative growth where temporary deficiencies occur
- In highly calcareous or highly acidic soils

Phosphorus (P) is taken up by plants primarily as phosphate ions ($H_2PO_4^-$ and HPO_4^{2-}) depending upon soil pH. In mineral soils, most phosphorus reacts to form fairly insoluble compounds. This is not necessarily a negative characteristic of phosphorus, because the limited solubility prevents leaching from the root zone. In neutral to alkaline soils, calcium phosphate compounds are formed, and in acid soils, iron and aluminum phosphates are produced. Figure 2-4 illustrates phosphorus fixation as affected by soil pH. Phosphorous availability is greatest in soils with a neutral pH.

Unlike soil, the properties of many soilless media frequently result in higher phosphorus solubility and mobility. As a result, the amount of added phosphorus needed for soilless media grown plants may be much less than for soil grown plants. However, since there are few mineral surface sites for phosphorus reaction in soilless media, the likelihood of P leaching can be significant, which can be an environmental concern. Phosphorus is frequently incorporated into a liquid fertilization program. However, sometimes pre-plant phosphorus is incorporated into the substrate before planting.

Plant-available soil phosphorus may be only 1 percent or less of the total phosphorus present in mineral soil, so its solubility is just

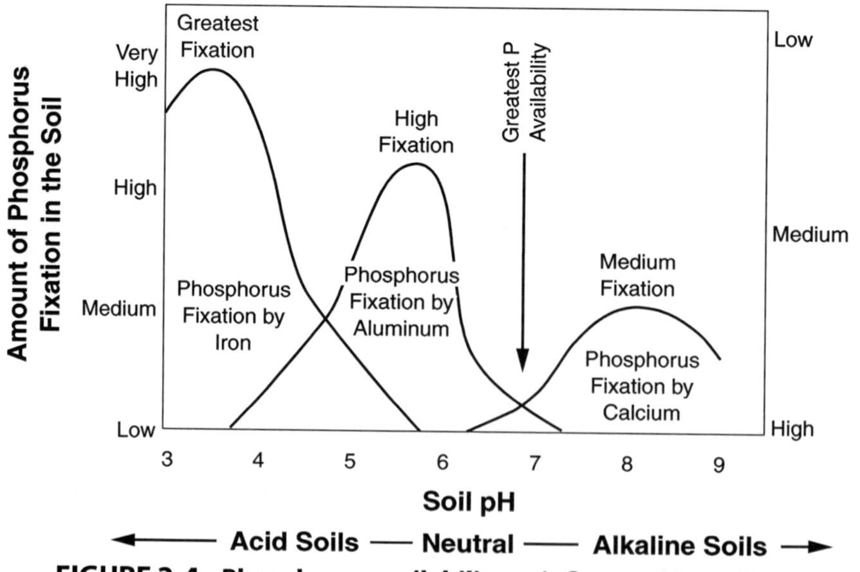

FIGURE 2-4. Phosphorus availability as influenced by soil pH.

as important as the total amount present. The solubility of phosphate in soil is controlled by several factors. The total amount of solid-phase (precipitated) phosphate in the soil is one factor. In general, the higher the total phosphorus content in the soil, the better the chance of having more phosphorus in solution. Another important factor is the extent of contact between solid-phase phosphate and the soil solution. As the exposure of phosphate to soil solution and to plant roots increases, so does the soil's ability to maintain replacement supplies. During periods of rapid plant growth, phosphorus in the soil solution may be replaced 10 times or more per day from solid-phase forms. Cool soil temperatures frequently induce phosphorus deficiency because of limited root growth and development, and lessened solubility. Maximum availability of soil phosphorus occurs at pH 6.0 to 7.0.

Research has demonstrated an increased uptake of phosphorus by some plants when the ammonium form of nitrogen is applied with phosphate fertilizer. This synergistic effect is maximized with banded or concentrated fertilizer applications. Various theories have been used to explain this observation. Some theories are: increased root growth; physiological changes making root cells more receptive to phosphorus; increased transfer of phosphorus across root membranes to the xylem; and lowering of soil pH from nitrification of ammonium nitrogen.

The amount of fertilizer applied should be based on a soil test, plant requirements, and/or expert advice. During turf or groundcover establishment, phosphorus, along with nitrogen and potassium, is normally incorporated into the soil. Subsequent periodic applications of phosphorus-containing fertilizers are common for rapidly growing turfgrasses. Mature ornamental trees, shrubs, and vines that have extensive root systems, and those grown in warm climates, generally require infrequent phosphorus addition. An exception is soils with a pH less than 6.0. When appropriate for trees, shrubs, and vines, phosphorus-containing fertilizers, including fertilizer spikes or tabs, are placed just below the soil surface around the outside margins of the root zone. Leguminous plants, such as *Acacia* and *Prosopis,* frequently require higher levels of phosphorus than other plant species.

Mycorrhizal fungi can have a particular benefit in phosphorus and zinc uptake. Ectotrophic mycorrhizal fungi are found on the roots of most tree species and may significantly improve plant growth. Vesicular arbuscular mycorrhizal fungi are found on the root systems of most other plants. Commercial sources of mycorrhizal fungi are available and may be beneficial for some situations.

A lack of sufficient phosphorus may or may not display visual symptoms in the plant. By the time phosphorous deficiency symptoms are apparent, a significant loss of growth and development has already occurred. This stresses the importance of tissue testing for the early detection of phosphorous deficiency. The use of deficiency symptoms as a guide for fertilization is not recommended. Depending on the plant species, symptoms of phosphorus deficiency in plants may include:

- Slow growth; stunted plants
- Purplish coloration on foliage of some plants
- Dark green coloration
- Delayed maturity
- Poor flower, fruit, or seed development

Potassium

Potassium (K) is taken up by plants in the form of potassium ions (K^+) and is highly mobile within plants. It is not synthesized into compounds, as are nitrogen and phosphorus, but always remains in this simple ionic form within cells and tissues. Potassium is essential for the translocation of sugars and for the formation of starch. More than 80 plant enzymes require K^+ for their activation. It is required in the opening and closing of stomata by guard cells, which is important for efficient water use. Potassium also promotes root growth; produces larger, more uniformly distributed xylem vessels throughout the root system; increases plant resistance to disease and environmental stress (e.g., drought); increases the size and quality of flowers; improves winter hardiness of perennials; and increases the durability of high traffic turf.

Soil minerals contain as much as 2 percent by weight of potassium. However, up to 98 percent of potassium is fixed in primary soil minerals and is unavailable to plants. From 1 to

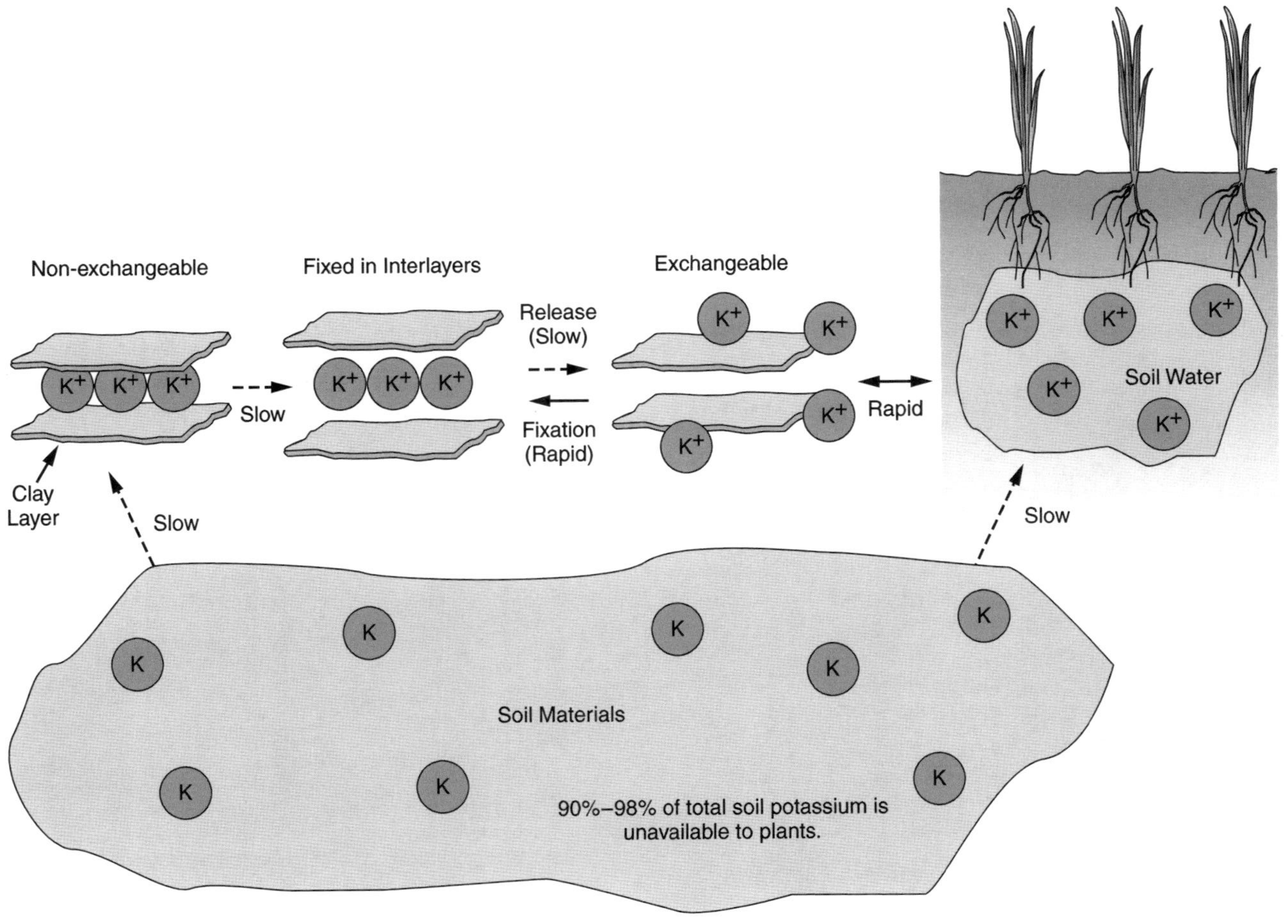

FIGURE 2-5. Availability of soil potassium.

10 percent of total soil potassium may be trapped in expanding lattice clays and is only slowly available. Between 1 and 2 percent potassium is contained in the soil solution and on exchange sites where it is readily available to plants. Figure 2-5 illustrates the various forms of potassium in soil. Soils that are likely deficient in potassium are acid, sandy, low in organic matter, and low in total cation exchange capacity. Soils high in calcium and/or magnesium have been reported to cause potassium deficiency in plants needing higher levels of potassium. Sandy soils and soilless mixes of low cation exchange capacity have a potential for potassium leaching.

Turfgrasses need for potassium is second only to its requirement for nitrogen. Large amounts of potassium are lost when turfgrass clippings are removed (Table 2-1).

Symptoms of potassium deficiency in plants include:

- Slow growth
- Tip and marginal chlorosis/necrosis starting on more mature leaves/needles
- Necrotic spots over whole leaf, generally starting with the lower leaves
- Weak stems and stalks
- Small fruit or shriveled seeds

TABLE 2-1
Nutrient Utilization by Various Turfgrasses
(Based on 50 lb clipping weight per 1,000 sq ft)

	N	P_2O_5	K_2O
Grass	**lb per 1,000 sq ft**		
Bentgrass	2.9	0.6	1.5
Bluegrass	0.8	0.4	0.8
Bermuda	1.3	0.3	1.2
Fescue	1.0	0.3	1.4
Ryegrass	1.1	0.5	1.2

SECONDARY PLANT NUTRIENTS

Plants require calcium, magnesium, and sulfur in amounts of similar magnitude to the primary nutrients. Since these nutrients are less frequently deficient, they are classified as secondary nutrients.

Calcium

Calcium (Ca) is taken up by plants as the calcium ion (Ca^{2+}). A structural nutrient, it is an essential part in cell walls and membranes and must be present for the formation of new cells. Calcium is important in the permeability of cell membranes. It is believed to counteract the toxic effects of oxalic acid by forming calcium oxalate in the vacuoles of cells. Calcium, once deposited in plant tissues, is not remobilized. Young tissue is affected first under conditions of deficiency. Calcium deficiency can cease plant growth due to the failure of terminal bud shoot development and apical tip root growth.

Most soils in the western United States contain more than enough calcium for plant use. One exception is very sandy, acidic soil, which has a low cation exchange capacity. It is too often assumed that calcium is so generally abundant that its only requirement as a fertilizer nutrient is on very acid soils where lime is needed. Almost all pre-packaged potting mixes contain a calcium source such as gypsum or lime.

Soils low in calcium and extremely acid soil conditions can result in deficiencies of other nutrients or toxicity of other elements such as aluminum and manganese. This can be more serious to plant health than the calcium deficiency alone. Limestone is added to acid soils to increase calcium availability along with increasing the soil pH. In alkaline soils where calcium solubility is low, supplemental calcium often is required. Ground limestone is routinely added to soilless media preparations since low calcium can be a concern. Calcium deficiency can be accentuated by high soil magnesium levels.

Calcium is immobile within plants so deficiency symptoms commonly appear on younger foliage. Symptoms of calcium deficiency in plants include:

- Tip burn of young leaves
- Death of growing points including terminal buds and root tips

- Abnormal, dark green appearance of foliage
- Premature shedding of blossoms and buds
- Weakened stems

Magnesium

Plant uptake of magnesium (Mg) is in the form of the magnesium ion (Mg^{2+}). The chlorophyll molecule, which is essential for photosynthesis, contains magnesium. Chlorophyll usually accounts for about 15 to 20 percent of the total magnesium content of plants. Magnesium serves as an activator of many plant enzymes required in growth processes.

The availability of magnesium is generally high in many western United States soils, but is more often deficient than calcium. Plants growing on sandy, acid soils may show magnesium deficiencies. High soil levels of calcium or potassium can induce a magnesium deficiency through competition for uptake. Applications of dolomitic limestone to acid soils low in both calcium and magnesium will add both of these nutrients to the soil.

Magnesium is mobile within plants and can be readily translocated from older to younger tissue when it is deficient. Symptoms of magnesium deficiency in plants include:

- Interveinal chlorosis in older leaves
- Curling of leaves upward along margins
- Marginal yellowing, with green "Christmas tree" area along mid-rib of leaf
- In conifers, orange-yellow and sometimes red needles, occurring in older needles first

Sulfur

Sulfur (S) is taken up from the soil primarily in the form of sulfate ion (SO_4^{2-}). Sulfur as sulfur dioxide (SO_2) in the air may also be absorbed through plant leaves. In recent years the atmospheric source of sulfur has been greatly reduced by a decrease in sulfur emissions.

Sulfur is an essential constituent of three amino acids (cystine, methionine, and cysteine) and is therefore necessary for protein synthesis. It is essential for nodule formation on legume roots. Sulfur is present in oil compounds responsible for the characteristic odors of plants such as garlic and onion.

Sulfur deficiencies are widespread throughout the United States and have been identified under a range of soil and climate conditions. Sulfur deficiencies are more likely found in soils low in organic matter. The sulfate ion is leachable, similar to nitrate, but to a lesser extent. Thus, plants grown in sandy soils with high rainfall may have sulfur deficiency. In low rainfall areas it is frequently found precipitated in the soil profile as gypsum ($CaSO_4 \bullet 2H_2O$). Sulfur deficiencies have become more common in recent years due to a number of factors, including higher air quality standards, increased plant production, and more use of relatively sulfur-free nitrogen and phosphorus sources. Some have been prompted to call sulfur the "fourth major nutrient."

Sulfur is generally immobile within plants. Symptoms of sulfur deficiency in plants include:

- Retarded growth rate and delayed maturity
- Young leaves light green to yellowish in color; in some plants, older tissue may also be affected
- In conifers, needles are similar to nitrogen deficiency—yellowish, short, and close together
- Small and spindly plants

MICRONUTRIENTS

Even though micronutrients are used by plants in very small amounts, they are just as essential for plant growth as the primary and secondary nutrients used in larger amounts. In ornamental plants and turf, specific micronutrient deficiencies are likely in western United States soils due to generally alkaline soil pH. In soilless media preparations, micronutrient fertilizers are normally added to ensure adequate amounts for plants.

Iron

Iron (Fe) is taken up by plants as ferrous (Fe^{2+}) or ferric (Fe^{3+}) ions. The chemical properties of iron make it important in oxidation-reduction reactions in plants. Iron is required for the formation of chlorophyll in plant cells. It serves as an activator for biochemical processes, such as respiration, photosynthesis, and symbiotic nitrogen fixation.

Most soil iron is unavailable to plants. Iron is a common micronutrient deficiency, which is most frequently observed in alkaline or calcareous soils. In calcareous soils, the deficiency is sometimes called lime-induced chlorosis. Iron deficiencies are also common in situations with poor soil aeration, excessive wetness, or when high levels of zinc, manganese, or copper are present. Turfgrass, certain trees, and ornamentals are especially susceptible to iron deficiency. Although iron deficiency is more likely in alkaline soils, sometimes deficiency occurs in sensitive plants, such as rhododendrons and camellias, even in acid and neutral soils.

Iron is immobile within plants. Symptoms of iron deficiency in plants include:

- Interveinal chlorosis of young leaves; veins remain green except in severe cases
- Twig dieback and in severe cases, death of entire limbs or plants
- In conifers, new growth is stunted and chlorotic

Zinc

Zinc (Zn) is taken up by plants as the zinc ion (Zn^{2+}) and is a structural component in enzymes involved in energy metabolism and synthesis of plant protein. It controls the synthesis of indoleacetic acid, an important plant hormone. Terminal growth areas are affected first when zinc is deficient.

Zinc is the micronutrient most often deficient in many western United States soils. Zinc deficiency is moderately common among cultivated trees and large shrubs, and its effect on growth can be significant. Foliar zinc sprays will temporarily correct deficiency in most plants. Soil applied or top dress applications of zinc may take longer to correct deficiency symptoms but may last longer than foliar applications. Many ornamental plants can be afflicted with zinc deficiency. Zinc deficiency is most common on alkaline pH soils or soilless media. Plants grown in soils that are sandy, have low organic matter content, or are calcareous are likely to be zinc deficient. Soils very high in available phosphorus may induce a zinc deficiency.

Zinc is mobile within plants. Symptoms of zinc deficiency in plants include:

- Decrease in stem length
- Rosetting of terminal leaves
- Reduced fruit bud formation
- Mottled young leaves or interveinal chlorosis
- Leaves are small (little leaf), very narrow, and pointed; older leaves drop
- Dieback of twigs after first year

Manganese

Manganese (Mn) uptake is in the form of the manganese ion (Mn^{2+}). Manganese serves as an activator for enzymes in plant growth processes. Manganese takes part in oxidation-reduction processes, and in decarboxylation and hydrolysis reactions. It assists iron in chlorophyll formation. High concentrations of manganese in plants may induce iron deficiency.

Landscape trees in calcareous soils may have manganese deficiencies. The availability of manganese for plants decreases as the soil pH increases above 7.0. High levels of copper, iron, or zinc can reduce manganese uptake by plants.

Manganese is immobile within plants. Symptoms of manganese deficiency in plants include:

- Interveinal chlorosis of young leaves
- Gradation of pale-green leaf coloration, with darker color next to veins; no sharp color distinction between veins and interveinal areas as with iron deficiency

Copper

Plant uptake of copper (Cu) is in the form of the copper ion (Cu^{2+}). Copper serves as an activator of several plant enzymes and may play a role in vitamin A production. A copper deficiency interferes with protein synthesis.

Native soil copper is usually adequate for normal plant growth. However, copper deficiency has been observed on trees, vines, and some annual plants growing in organic soils and sands. Copper can be toxic at low levels. Therefore, copper application is not recommended except where the need has been established through tissue and soil testing.

Copper is immobile within plants. Symptoms of copper deficiency in plants include:

- Stunted growth
- Dieback of terminal shoots in trees
- Poor pigmentation
- Wilting and eventual death of leaf tips

Boron

Boron (B) is predominately in solution as boric acid [H_3BO_3 or $B(OH)_3$] and is taken up by plants and transported in the xylem in this form in soils below pH 9. It functions in plants in the differentiation of meristematic cells. When boron is deficient, cells may continue to divide, but structural components are not differentiated. Boron is also involved in regulating metabolism of carbohydrates in plants. Leafy plants require more boron than grasses. Plant reproductive organs require more boron than vegetative tissues. Foliar-applied boron may be transported through the phloem of some, but not all, plants. It has been determined that boron is immobile in certain plant species. Little is known about boron mobility in ornamental plants.

Symptoms of boron deficiency in plants include:

- Death of terminal growth, causing lateral buds to develop, producing a “witch’s broom” effect
- Thickened, curled, wilted, and chlorotic leaves
- Soft or necrotic spots on corms
- Reduced flowering or improper pollination

Boron toxicity in ornamentals can occur. Toxicities are usually associated with high levels of boron in irrigation water. Care must be exercised in the use of boron fertilizer, since the difference between deficient and excessive levels is often small.

Molybdenum

Molybdenum (Mo) is taken up by plants as the molybdate ion (MoO_4^{2-}). Plants require molybdenum for the proper utilization of nitrogen. Nitrate is converted to amino acids by the nitrate-reductase enzyme, which requires molybdenum. In addition, molybdenum is required for the symbiotic fixation of nitrogen by legumes.

Molybdenum deficiency in woody plants is rare. Poinsettia is one plant reported to have the potential of a molybdenum deficiency. Since molybdenum is critical in nitrogen fixation, leguminous trees or shrubs such as *Acacia*, alder, and locust may have low soil molybdenum, which causes plants to exhibit nitrogen deficiency.

Molybdenum is mobile within plants. Symptoms of molybdenum deficiency in plants include:

- Stunting and lack of vigor
- Marginal scorching, cupping, or rolling of leaves

Chlorine

Chlorine (Cl) is taken up by plants as the chloride ion (Cl^-) and it is mobile within plants. It is required in photosynthetic reactions in plants where water is split and oxygen released. It is also utilized in regulating cell turgor pressure. Until the mid-1980s, deficiencies of chlorine in the field were believed to be rare. However, chloride-containing fertilizers are now being used specifically to increase disease resistance of many plants. Some nutritional benefits have also been reported. Chloride toxicity may be a concern for many horticultural plants.

Nickel

Nickel (Ni) is absorbed by plants as the nickel ion (Ni^{2+}). It is a recent addition to the list of essential plant nutrients, having been recognized as such only since 1989. Nickel is a component of the enzyme urease, which is necessary for the conversion of urea to ammonia in plant tissue. It is, therefore, important to overall nitrogen metabolism.

Nickel is required in very small amounts. The critical level in plant tissue appears to be around 0.1 ppm. There are very few deficiencies of nickel in the West, so nickel deficiency is more an academic than commercial interest at this time.

Nickel is considered immobile within plants. Deficiency symptoms produced under research conditions include:

- Chlorosis of young leaves
- Death of the meristem

NUTRIENT INTERACTIONS

One should not focus on just one element when thinking about plant nutrition. Plant nutrients interact in a complex web of bio- and geochemical reactions. Some plant nutrients interact more with certain nutrients than others. The following section explores some concepts that describe principles of plant nutrient interactions.

Plants must maintain electrical neutrality. Thus, a plant root will absorb as many positive (cation) electrical charges as it will the negative (anion) electrical charges. For example, plants absorb approximately equal amounts of K^+ electrical charges as the plant absorbs of NO_3^- electrical charges. This leads to an important principle: cations compete with other cations and anions compete with other anions for absorption by plant roots.

Macronutrient Interactions

In Tables 2-2 and 2-3 the interactions among nutrients will be grouped into macronutrients and micronutrients. Nitrogen will be split between ammonium (NH_4^+) and nitrate (NO_3^-). The two tables of macronutrient and micronutrient interactions apply specifically to nutrient absorption by plant roots in soil.

Micronutrient Interactions

Several macronutrients have consistent interactions with micronutrients. Application of urea or ammonium fertilizer results in

TABLE 2-2

Macronutrient Interactions

Nutrient	Ammonium NH_4^+	Potassium K^+	Calcium Ca^{2+}	Magnesium Mg^{2+}	Nitrate NO_3^-	Phosphate $H_2PO_4^-$	Sulfate SO_4^{2-}
Ammonium	—	D	D	D	I	I	I
Potassium	D	—	D	D	I	I	I
Calcium	D	D	—	D	I	I	I
Magnesium	D	D	D	—	I	I	I
Nitrate	I	I	I	I	—	D	D
Phosphate	I	I	?	?	D	—	D
Sulfate	I	I	I	I	D	D	—

Thomas Ruehr, Ph.D., California Polytechnic University, San Luis Obispo.

(I) indicates adding the nutrient on the left will increase the plant absorption of the nutrient in the top row

(D) indicates the nutrient on the left will decrease the absorption of the nutrient in the top row

(?) indicates insufficient information is available to be certain of the interaction, or the interaction is more dependent upon plant type or the specific soil conditions

the production of acidity on a microscopic scale in soil. The acid dissolves some of the native micronutrient metal ions (iron, manganese, copper, zinc, and nickel) increasing the availability to the plant.

A high level of soil phosphate causes the phosphate ion to enter the plant roots. Inside the roots, this phosphate inhibits the movement of the same micronutrient metal ions from translocating upward to the plant shoots.

Sulfate (SO_4^{2-}) in high concentration suppresses the plant root's ability to absorb molybdate (MoO_4^{2-}) from the soil because the ion forms are nearly identical. Thus, heavy application of gypsum or ammonium sulfate can create a molybdenum deficiency. The symptoms would resemble an overall nitrogen deficiency due to the role molybdenum plays in almost all nitrogen reactions.

TABLE 2-3
Micronutrient Interactions

Nutrient	Iron Fe^{2+}	Manganese Mn^{2+}	Copper Cu^{2+}	Zinc Zn^{2+}	Nickel Ni^{2+}	Chloride Cl^{-}	Molybdate MoO_4^{2-}	Boric acid H_3BO_3
Iron	—	D	D	D	D	I	I	?
Manganese	D	—	D	D	D	I	I	?
Copper	D	D	—	D	D	I	I	?
Zinc	D	D	D	—	D	I	I	?
Nickel	D	D	D	D	—	I	I	?
Chloride	I	I	I	I	I	—	D	?
Molybdate	I	I	I	I	I	D	—	?
Boric acid	I	I	I	I	I	D	D	—

Thomas Ruehr, Ph.D., California Polytechnic University, San Luis Obispo.

(I) indicates adding the nutrient on the left will increase the plant absorption of the nutrient in the top row

(D) indicates the nutrient on the left will decrease the absorption of the nutrient in the top row

(?) indicates insufficient information is available to be certain of the interaction, or the interaction is more dependent upon plant type or the specific soil conditions

DIAGNOSING NUTRIENT NEEDS

Visual symptoms of nutrient deficiencies or toxicities can be a useful tool for diagnosing problems. However, assistance should be obtained from a qualified person, since chlorosis (yellowing) and necrosis (death) of tissues can result from non-nutritional problems such as diseases and insects. In addition, soil and other rooting media testing are very useful in identifying nutrient deficiencies, especially in regards to primary and secondary nutrients. Where appropriate, plant tissue testing can help confirm a specific deficiency of any nutrient. Finally, toxicity from excessive amounts of certain elements, damage from herbicides, or lack of proper aeration in the soil root zone can produce yellowing or death of plant tissue. The diagnosis of nutrient needs is covered more completely in Chapter 11.

SUPPLEMENTARY READING

Bennett, W.F., ed. 1993. *Nutrient Deficiencies & Toxicities in Crop Plants.* The American Phytopathological Society, St. Paul, MN.

Follett, R.F., D.R. Keeney, and R.M. Cruse, ed. 1991. *Managing Nitrogen for Groundwater Quality and Farm Profitability.* America Society of Agronomy, Madison, WI.

Hall, K.D., and K.E. Nowels, 2010. *Fertilizer 101*. The Fertilizer Institute, Washington, D.C.

Hauck, R.D. ed. 1984. *Nitrogen in Crop Production.* American Society of Agronomy, Madison, WI.

Khasawneh, F.E., ed. 1980. *The Role of Phosphorus in Agriculture.* American Society of Agronomy, Madison, WI.

Marschner, H. 1995. *Mineral Nutrition of Higher Plants*. Academic Press, San Diego, CA.

Mortvedt, J.J., F.R. Cox, L.M. Shuman, and R.M. Welch, eds. 1991. *Micronutrients in Agriculture*, 2nd ed. Soil Science Society of America, Madison, WI.

Munson, R.D., ed. 1985. *Potassium in Agriculture*. American Society of Agronomy, Madison, WI.

Olson, R.A., and K.J. Frey, ed. 1987. *Sulfur in Agriculture*. American Society of Agronomy, Madison, WI.

Schroeder, C.B. 1997. *Introduction to Horticulture, Science & Technology*, 2nd ed. Interstate Publishers, Inc. Danville, IL.

Tisdale, S.L., W.L. Nelson, J.D. Beaton, and J.L. Havlin. 1993. *Soil Fertility and Fertilizers*, 5th ed. The Macmillan Company, New York, NY.

CHAPTER 3

Soils

Soil is a critical component of healthy and productive landscapes. Soil may also be part of artificial growing media mixes. A good understanding of the basic principles of soil properties, chemistry, and biology will aid in proper care and management of turf and ornamental landscapes. Soilless media are discussed in detail in Chapter 4.

WHAT IS SOIL?

Soil can be described as a complex natural material derived from disintegrated, decomposed, and reformed minerals and organic matter that provides nutrients, moisture, and anchorage for land plants. For agricultural purposes, soils are generally modified only slightly from their natural condition. In contrast, for many turf and ornamental situations, natural soil may be extensively modified. Most potting media used in greenhouses and nurseries contain little, if any, soil.

The four principal components of soil are minerals, organic matter, water, and air. These are combined in widely varying amounts in different kinds of soil and at different moisture levels. Soils typically are 50 percent solid materials and 50 percent pore space on a volume basis. The pore space contains approximately equal volumes of water and air as shown in Figure 3-1.

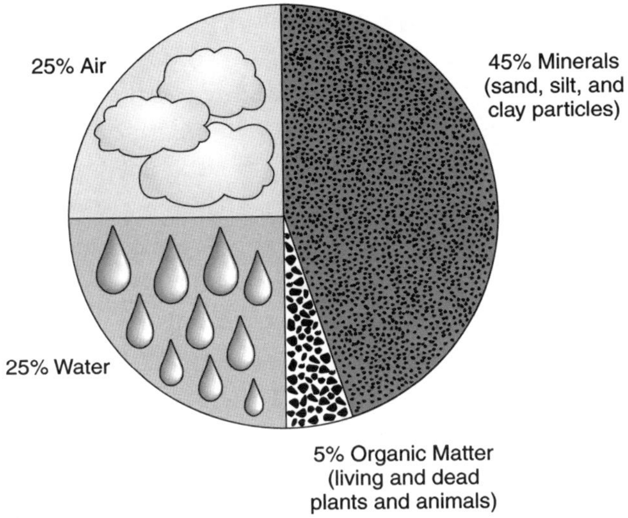

FIGURE 3-1. Volumetric content of four principal soil components for a representative soil at an ideal moisture content for plant growth.

HOW SOILS ARE FORMED

The development of soils from rocks and minerals is a long-term process that is called *weathering*. Weathering reduces rocks and minerals to smaller particles that eventually become soil. Weathering takes place without major movement of rocks and minerals as compared to *erosion* where rocks are broken as they are moved by various forces. The two main types of weathering are physical and chemical. Physical and chemical weathering can take place at the same time.

Physical weathering breaks rock down mainly by mechanical abrasion. However, direct exposure of rocks to the environmental processes such as heat, frost, water abrasion, pressure release, and salt formations are also important forms of physical weathering.

Chemical weathering takes place by action of water, oxygen, carbon dioxide, and naturally occurring acids to reduce the size of rock fragments and change the chemical composition of many of the resulting particles.

Biological and biochemical weathering are also an important means of reducing rocks and minerals to smaller sizes. Chemicals produced by microorganisms and higher plant and animal life can degrade the minerals they contact. Plant roots can slowly break apart rocks. The remains of plant and animal life contribute organic matter to the weathered rock material and a true soil begins to form.

In some cases human activity has drastically changed soil from its original condition. Examples of such activities include land leveling, vehicle traffic, and subsurface tillage carried out to break up hardpan layers. Golf courses and landscapes may consist of "soil" that is constructed from materials that were transported great distances from their origin.

Hundreds or thousands of years are required for the formation of just one inch of soil. The widely variable characteristics of soils are due to differential influences of five soil-forming factors:

- Parent material — material from which soils were formed
- Climate — temperature and moisture
- Living organisms — microscopic and macroscopic plants and animals
- Topography — shape and position of land surfaces
- Time — period during which parent materials have been subjected to weathering processes

While soil properties can be very quickly altered by events such as landslides, flooding, or machinery, the process of soil formation is slow. Evidence indicates that the soils on which we depend required centuries to form and, therefore, soil should be considered a non-renewable resource. Thus, it is important that soils are protected from erosive forces and nutrient depletion, which can rapidly destroy the product of many years of development.

SOIL PROFILE

A vertical section through a soil typically presents a layered pattern. The vertical section is called a *profile.* Each individual layer is referred to as a *horizon.* An example of a soil profile with three horizons is shown in Figure 3-2.

The uppermost horizon includes the *surface soil,* or *topsoil,* and is designated the *A horizon.* The next successive horizon,

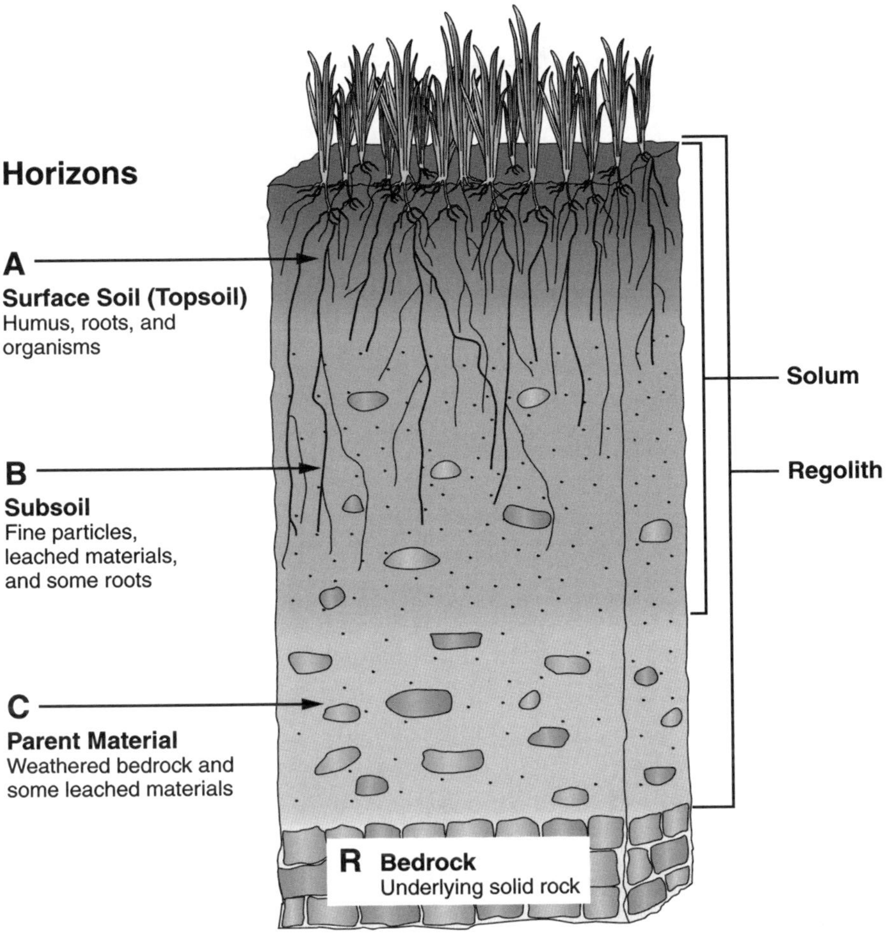

FIGURE 3-2. The typical soil profile illustrates the three general soil horizons.

underlying the surface soil, is called the *subsoil,* or the *B horizon.* Underlying the B horizon is the *parent material,* or the *C horizon.* The bedrock is beneath the parent material and is sometimes denoted with the letter R.

Soil profiles vary greatly in depth or thickness, from a fraction of an inch to many feet. Normally, however, a soil profile will extend to a depth of about three to six feet. Other soil characteristics, such as color, texture, structure, and chemical composition, also exhibit wide variations among the many soil types.

The surface soil (A horizon) is the layer most subject to climatic and biological influences. Most of the organic matter accumulates in the surface layer, which usually gives it a darker color than the underlying horizons. If enough time has passed and in higher rainfall climates, this layer is characterized by a loss of soluble and colloidal materials that are moved into the lower horizons by infiltrating water, a process called *eluviation* or *leaching.*

The subsoil (B horizon) is the layer that commonly accumulates many of the materials leached and transported from the surface soil. This accumulation is termed *illuviation.* Development of a B horizon takes many years. The western United States has many acres of productive soils that are too young to have developed a B horizon.

The deposition of materials such as clay particles, iron, aluminum, calcium carbonate, calcium sulfate, and other salts creates a layer that normally has a more compact structure than the surface soil. In severe cases, this deposition may lead to the formation of cemented layers. This compact structure often leads to restricted movement of moisture and air, which can limit root growth. Cemented layers formed from illuviation of calcium carbonate are called *caliche.* Cemented layers formed from clay particles are referred to as a *claypan.*

The parent material (C horizon) is the least affected by physical, chemical, and biological agents. It is more similar in chemical composition to the original material from which the A and B horizons were formed. Parent material that has formed in its original position by weathering of bedrock is termed *sedentary* or *residual.* Parent material that has been moved to a new location by natural forces is called *transported.* Transported parent material is further characterized on the basis of the kind of natural force responsible for its transportation and deposition. When water is the transporting agent, the parent materials are referred to as *alluvial* (stream deposited), *marine* (sea deposited), or *lacustrine* (lake deposited). Wind-deposited material is called *aeolian.* Material transported by gravity is called *colluvial,* a category that is associated with mudslides and hillside soil creep.

Because of the strong influence of climate on soil profile development, certain general characteristics of soils formed in areas of different climatic patterns can be described. For example, much of the western United States has an arid or semi-arid climate that results in the development of soils that are coarser in texture than many of those developed in more humid climates. Also, many western soil profiles are less developed than those in the eastern United States, since the amount of water percolating through the soil is generally much less than in more humid climates. Many western soils, therefore, contain more calcium, potassium, phosphorus, and other nutrients than extensively weathered eastern soils with higher rainfall and intensive leaching. However, highly weathered soils can be found in high rainfall areas of the Pacific Northwest and northwestern California.

Thus, the soil profile is an important consideration when growing plants. To a large extent the depth of soil, its texture and structure, and its chemical nature determine the value of the soil as a medium for plant growth.

SOIL TEXTURE

Soils are composed of particles with a wide range of sizes and shapes. Individual mineral particles are divided into three categories, known as *soil separates*, on the basis of their size: sand, silt, and clay (Figure 3-3). Such a division is meaningful, not only in terms of a classification system, but also in relation to plant growth. Many important chemical and physical reactions occur on the surface of soil particles. The relative surface area increases as particle

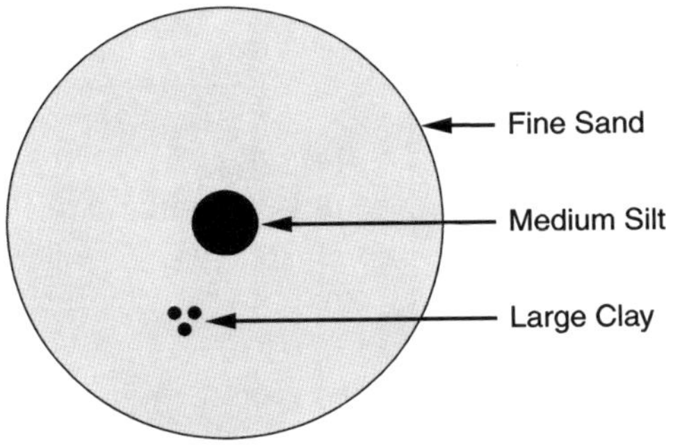

FIGURE 3-3. Relative size of sand, silt, and clay.

size diminishes, which means that the smallest particles (clay) are the most important for chemical and biological reactions.

The size ranges for sand, silt, and clay in the USDA and the International Classification systems are shown in Table 3-1. Soil texture descriptions are based on the proportions of sand, silt, and clay. Twelve basic soil textural classes are recognized (Figure 3-4).

A textural class description of soils reveals much about potential soil-plant interactions. Many of the physical and chemical properties of soils are determined by texture. For example, in a mineral soil, the cation exchange capacity is closely related to the amount and type of clay in the soil. The water-holding capacity is determined in large measure by the particle size distribution. Fine-textured soils (high percentage of silt and clay) hold more water than coarse-textured soils (sandy). Fine-textured soils often are more easily compacted when wet, have slower movement of water and air, and require more energy to till and level. The need for less compact soils that can withstand high traffic and drain quickly after rain or irrigation is the reason most golf courses and sports fields are developed on sandy soils.

Medium-textured soils, such as loams, sandy loams, and silt loams, are usually the easiest soils in which to manage plant

TABLE 3-1

Size Limits of Soil Separates

U.S. Department of Agriculture			International	
Name of Separate	Diameter		Diameter	
	(Range)	(Sieve No.)	Fraction	(Range)
	(mm)			*(mm)*
Very coarse sand	2.00 – 1.00	10 – 18	I	2.00 – 0.20
Coarse sand	1.00 – 0.50	18 – 35	II	0.20 – 0.02
Medium sand	0.50 – 0.25	35 – 60	III	0.02 – 0.002
Fine sand	0.25 – 0.10	60 – 140	IV	Below 0.002
Very fine sand	0.10 – 0.05	140 – 270		
Silt	0.05 – 0.002	270 – 635		
Clay	Below 0.002	>635		

Soil Survey Manual. USDA Agricultural Handbook No. 18.

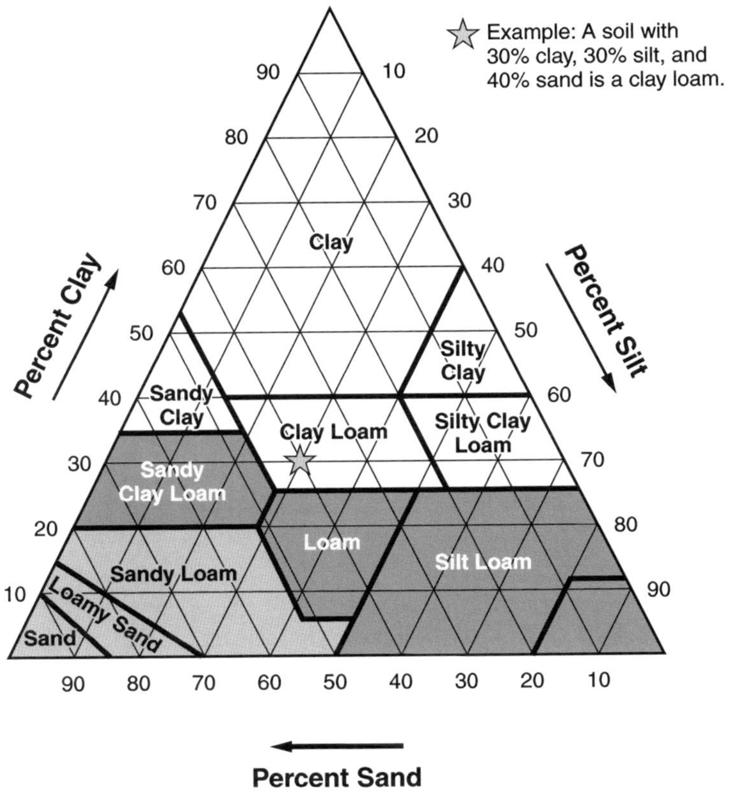

FIGURE 3-4. Soil textural triangle showing the percentages of sand, silt and clay in the basic textural classes.

growth. Nevertheless, there is only a general relationship between soil textural class and soil productivity because texture is only one of many factors that influence plant growth.

SOIL STRUCTURE

The way in which particles are grouped together is termed *soil structure.* Soil particles do not normally exist individually in the soil. Soils are normally arranged into aggregates or groups of particles. One exception is sand, which may exist as single grains.

There are four primary types of structure, based on shape and arrangement of the aggregates (Figure 3-5). When the particles are

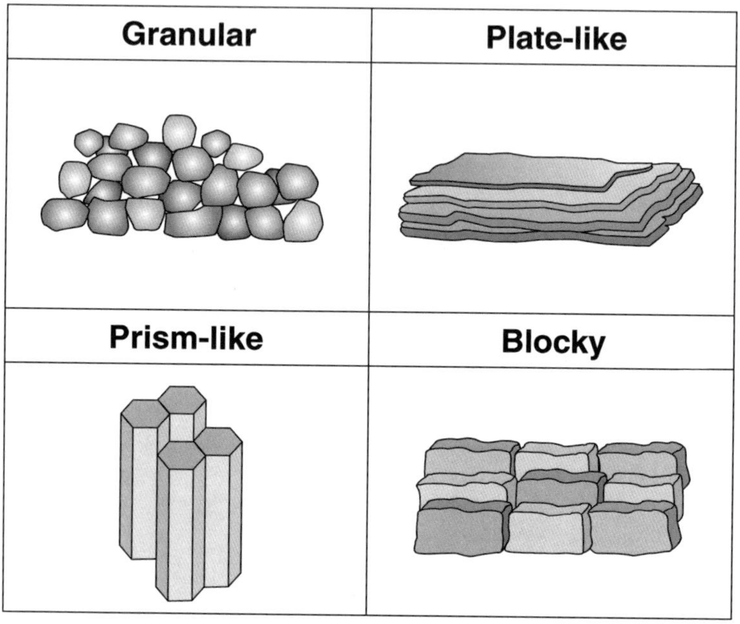

FIGURE 3-5. Generalized illustration of soil structural aggregates.

arranged around a horizontal plane, the structure is called *platy.* Platy structure can occur in any part of the profile. Puddling or ponding of soils often results in platy structure on the soil surface. When particles are arranged around a vertical line, bounded by relatively flat vertical surfaces, the structure is referred to as *prism-like,* with sub-types of prismatic and columnar. Prism-like structures are usually found in subsoils and are common in arid and semi-arid regions. Prismatic structure is prism-like with an angular level top. Columnar structure is prism-like with a rounded top and is commonly found in soils affected by excess sodium. The next type of structure is referred to as *blocky* and is characterized by approximately equal length in all three dimensions. This arrangement of aggregates is common in subsoils, particularly those in humid regions. Angular blocky is a block-like structure that has sharp, well-defined edges and rectangular faces. A subangular blocky structure is block-like but edges are somewhat rounded. Finally, *granular* structure includes all rounded aggregates. Granular structure is characteristic of many surface soils, particularly where the organic matter content is high. Massive and single-grain conditions are considered to have no structure. Soil management practices have an important influence on structure.

Soil aggregates are formed by both physical forces and binding agents. Binding agents are derived from soil microorganisms, plant roots, and the products of organic matter decomposition. The aggregates formed by binding agents are more stable and are more resistant to the destructive forces of water and cultivation. Conversely, aggregates formed by physical forces, such as drying, freezing and thawing, and tillage operations, are relatively unstable and more easily degraded. Excessive sodium also has a damaging effect on soil aggregation.

Soil structure has a strong influence on plant growth. Soil structure affects aeration, heat transfer, moisture relationships, water movement, and mechanical impedance of root growth. For example, good seedbed preparation is important for moisture and

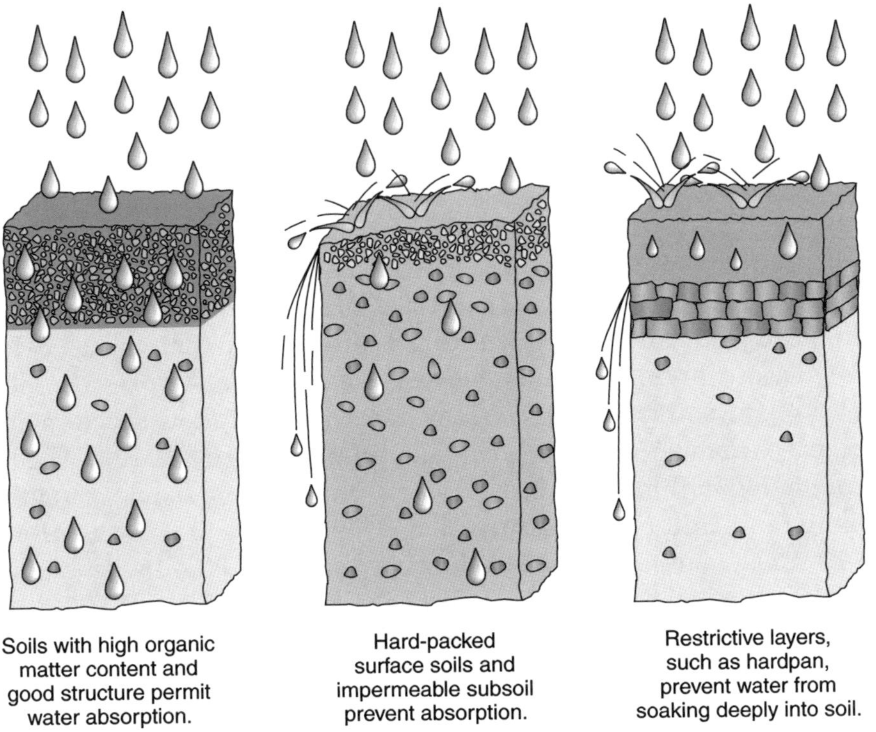

FIGURE 3-6. Soil permeability is dependent upon soil structure.

heat transfer, both of which affect seed germination. A fine, granular structure is usually ideal.

The movement of moisture and gases through the soil is dependent upon porosity, which is markedly influenced by soil structure. Granular structure provides adequate porosity for good water infiltration and gas exchange between soil and the atmosphere. This condition creates an ideal physical medium for plant growth. However, where surface crusting occurs or subsurface restrictive layers exist, plant growth is hindered because of reduced porosity. Good management practices improve soil structure by creating a better condition for plant growth. Conversely, excessive tillage and traffic when the soil is wet can destroy soil structure and cause compaction.

SOIL REACTION (pH)

Soil reaction (acid, neutral, alkaline) refers to the relative concentration of hydrogen ions (H^+) and hydroxyl ions (OH^-) in the soil solution. An acid soil has a higher concentration of hydrogen ions (H^+) than hydroxyl ions. Alkaline soil has more hydroxyl ions (OH^-) than hydrogen ions. A neutral soil has equal concentrations of hydrogen ions and hydroxyl ions.

The actual concentrations of hydrogen or hydroxyl ions are extremely low. A neutral soil has only 0.0000001 or 10^{-7} moles of hydrogen ions per liter of soil solution. This is a cumbersome way of reporting soil acidity, so the term *pH* is used instead. The pH scale ranges from 0 to 14, where pH 7 is neutral, pH values below 7 are acidic, and pH values above 7 are alkaline. As pH decreases below 7, the soil is considered to be more acidic. Likewise, as the pH increases above 7, the soil is considered to be more alkaline. The pH scale is logarithmic, so each pH unit change represents a tenfold change in acidity or alkalinity. For example, a soil with a pH of 5 is 10 times more acidic than one with a pH of 6. A soil with a pH of 8 is 10 times more alkaline than one with a pH of 7 and 100 times more alkaline than one with a pH of 6.

Soil pH is an important property since it influences many factors including nutrient availability, solubility of toxic ions, and microbial activity. Nutrient availability in soil varies with pH values due to chemical reactions. The maximum availability of many nutrients in mineral soil is at pH values between 6.5 and 7.5 (Figure 3-7). The solubility of some metal micronutrients and some potentially phytotoxic elements, such as aluminum and manganese, increases at low pH values (below 5.5), which can reduce plant growth. At pH above 7, the availability of many essential micronutrients becomes limited. Near neutral pH values favor the activity of many microorganisms responsible for essential soil biological activity.

There are five sources of soil acidity:

- Basic cations, calcium, magnesium, sodium, and potassium are leached from the topsoil into the subsoil as a result of rain and irrigation, leaving behind acidic hydrogen and aluminum ions
- Basic cations are taken up by plants and removed during plant removal

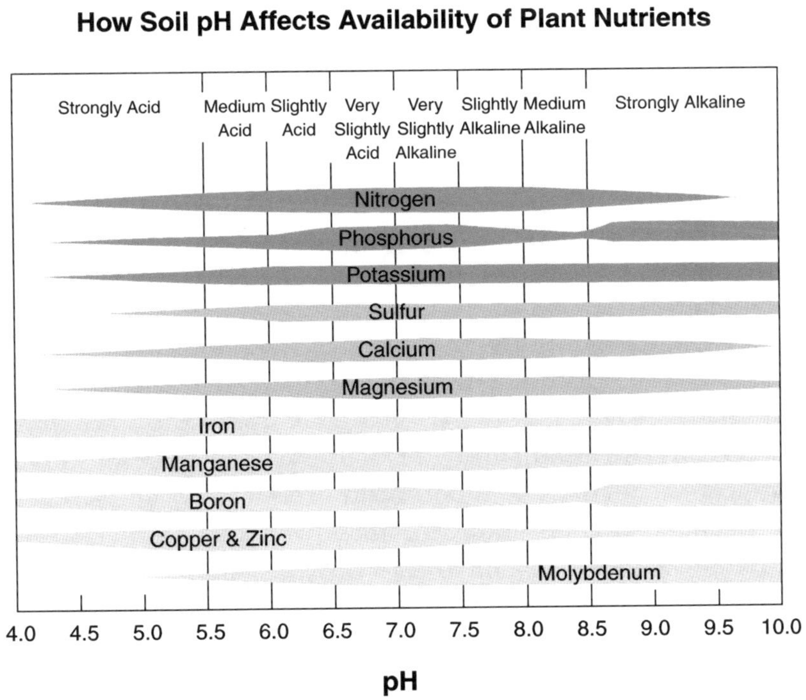

FIGURE 3-7. Relative mineral nutrient availability in mineral soil as affected by soil pH.

- The use of nitrogen fertilizers containing urea and ammonium lowers soil pH through the process of nitrification
- The use of elemental sulfur or various acids for soil reclamation or plant nutrition lowers soil pH
- Biological processes such as protons (H^+ ions) excreted by plant roots, carbonic acid derived from decomposition of humus and root respiration, and acid rain

The five sources of acidity are independent of each other and may occur separately or concurrently.

Calcium carbonate (free lime) found in many western arid soils acts as a buffer against the development of acidity. This buffering occurs because the solubility of calcium carbonate is increased as acidity increases, raising the level of exchangeable calcium in the soil. Soluble calcium replaces hydrogen ions on the exchange sites. The hydrogen ions then react with the carbonate to form water and carbon dioxide. The presence of calcium carbonate, combined with low rainfall, explains why many western soils are alkaline. However, the long-term use of acid-forming materials, mainly ammonium-based nitrogen fertilizers, has increased the incidence of acid soils in the western U.S., especially where soils initially lacked calcium carbonate.

Additions of calcium carbonate and other liming materials can be used to raise the pH of acid soils through the same mechanism. See Chapter 6 for more information on correcting acid soils with amendments.

CATION EXCHANGE CAPACITY

Cation exchange capacity (CEC) is a measure of the total quantity of cations that can be adsorbed or held by a soil. It is an important indicator of the fertility and potential productivity of a soil.

Because of their chemical structure, clay particles and soil organic matter generally have a net negative charge. This means that cations (positively charged ions) can be attracted to and held on the surface of these soil materials (attraction of opposites). Cations in the soil solution are in dynamic equilibrium with the cations adsorbed on the surface of the clay and organic matter. This exchange is shown in Figure 3-8. Soil organic matter generally has

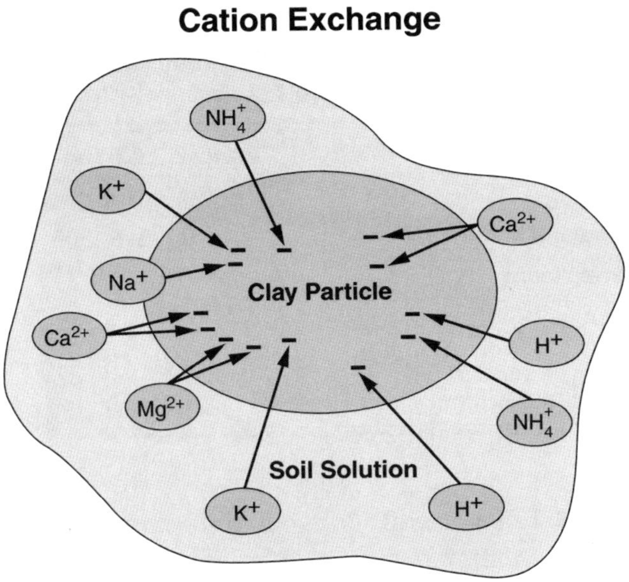

FIGURE 3-8. Schematic illustration of the exchange of cations between the negatively charged clay particles and the soil solution.

a negative charge and a large cation exchange capacity. However, the source of negative charge from organic matter is different than from clays. The negative charge in clays occurs during the crystallization of the minerals, while a negative charge in organic matter depends on the soil pH (variable or pH-dependent charge). The CEC of organic matter increases as the pH increases and the CEC decreases as the pH drops. A pH-dependant CEC becomes an important management factor when working with peat, muck, or soilless media composed mainly of organic matter.

The cations present in the largest quantities with respect to plant growth are calcium (Ca^{2+}), magnesium (Mg^{2+}), potassium (K^+), ammonium (NH_4^+), sodium (Na^+), and hydrogen (H^+). Positively charged micronutrient metals such as zinc (Zn^{2+}), copper (Cu^{2+}), iron (Fe^{2+} and Fe^{3+}), manganese (Mn^{2+}), and nickel (Ni^{2+}) are important for plant growth but are present at low concentrations. The relative amount of each cation adsorbed on the surface of clay particles is closely associated with important soil properties. Highly acid soils have a high percentage of adsorbed hydrogen, while soils with a

TABLE 3-2

Typical Cation Exchange Capacity of Some Soil Texture Classes

Soil Texture	Typical CEC Range
	(meq/100 g)
Sand and loamy sand	2–6
Sandy loam	3–8
Loam	7–15
Silt loam	10–18
Clay and clay loam	15–30

favorable pH of 6 to 8 generally have a high percentage of adsorbed calcium ions. Soils high in exchangeable calcium ions are more likely to be well aggregated and have higher infiltration rates. Soils that are high in sodium ions will have a pH greater than 8 and are dispersed while resisting water infiltration.

In the western United States, mineral soils with high CECs tend to be more fertile and easier to manage than those with low CECs. In a high CEC soil, nutrients such as potassium, calcium, and magnesium are less likely to be lost by leaching and such soils have a greater ability to store and supply nutrients to plants. The approximate CEC range for some typical soil texture classes is shown in Table 3-2. The range of values within a textural class is partly due to differences in organic matter content and clay mineralogy.

SOIL ORGANIC MATTER

Soil organic matter is one of the major contributors to soil productivity and fertility. Its importance to plant growth cannot be overstated. Soil organic matter:

- Helps strengthen soil aggregates, thus improving soil structure
- Improves aeration and water infiltration
- Increases water-holding capacity

- Provides significant amounts of cation exchange capacity
- Provides buffering against rapid changes in soil pH when acid- or alkaline-forming materials are added to soil
- Forms stable organic compounds that can increase the availability of micronutrients
- Serves as a slow-release source of many plant nutrients
- Serves as an energy source for soil microorganisms

Western soils in arid climates typically contain lower amounts of organic matter than soils formed in cooler, higher-rainfall areas. Many western soils have less than 1 to 2 percent organic matter, while soils in the Midwestern United States may have as much as 7 to 10 percent organic matter. Some western soils, however, such as those formed in river deltas or old lakebeds, may have much higher levels of organic matter.

Soil organic matter consists of plant and animal residues in various stages of decomposition, living organisms, and the substances synthesized by these organisms. Humus is the well decomposed, more or less stable part of the organic matter in mineral soils. Humus is composed of humic substances, which are complex compounds that resist decomposition. Humic substances are a complex mixture of many organic compounds including lignin, amino acids, carbohydrates, hemicellulose, cellulose, lipids, and many other compounds (Table 3-3). Non-humic substances are simple compounds that are rapidly decomposed by soil organisms, mainly bacteria, actinomycetes, and fungi. Non-humic substances are lipids, carbohydrates, proteins, and amino acids. Humic substances make up more than 60 percent of soil organic carbon.

Soil organisms play a major role in the decomposition of organic matter. Each group of organisms becomes a dominant factor at various times as decomposition proceeds. These microorganisms also excrete diverse compounds (such as polysaccharides) that can modify soil chemical and physical properties, such as aggregating soil particles together. Larger organisms, such as earthworms and insects, also ingest organic residue and soil particles, thereby binding them together into stable aggregates.

Organic matter serves as a major reservoir of nitrogen, phosphorus, and sulfur for plant nutrition. In the process of *mineralization*,

TABLE 3-3
Common Constituents of Soil Organic Matter and Relative Rates of Decomposition

Organic Constituents	Approximate Percentage of Total Organic Matter	Rate of Decomposition
Sugars, starches, simple proteins	1–5	Rapidly decomposed
Crude proteins	5–20	↑
Hemicelluloses	10–25	↕
Cellulose	30–50	↓
Lignins, fats, waxes	10–30	Very slowly decomposed

soil fungi and bacteria break down organic matter, converting complex carbon compounds to CO_2 and releasing nutrients in inorganic, plant-available forms, as shown in the following general reaction:

$$\text{organic matter} + H_2O + O_2 \xrightarrow{\text{soil organisms}} CO_2 + H_2O + NH_4^+ + H_3PO_4 + H_2S + \text{other gases} + \text{energy}$$

The amount of nutrient converted to plant-available form annually by mineralization depends mainly on the quantity of "active" organic matter (i.e., partially decomposed, younger organic residues) rather than the total organic matter that includes the relatively stable humus. However, a very general rule of thumb is that 30–60 pounds of N per acre (0.7 to 1.4 pounds per 1,000 square feet) are released for each percentage point of soil organic matter content. In addition, decomposition of organic matter will release smaller amounts of phosphorus and sulfur.

The soil organisms responsible for decomposing organic matter require nutrients for their growth and metabolism. The

TABLE 3-4
Approximate C:N Ratios of Common Materials

Material	Approximate C:N ratio
Fish meal	4
Grass clippings	17
Cow manure	18
Yard waste	23
Leaves	48
Paper	170
Sawdust	500

Modified from: Stoffella, P.J., and B.A. Kahn. 2001. "Compost Utilization in Horticultural Cropping Systems." *CRC Press.* Table 2.3, page 27.

carbon:nitrogen ratio of stable soil organic matter is usually assumed to be 10:1. The process of decomposing organic material with a high carbon:nitrogen ratio will temporarily tie up available nitrogen, reducing its availability to plants. This process is called *immobilization*. Many plant residues commonly returned to the soil have carbon:nitrogen ratios greater than 20:1 and will immobilize soil nitrogen unless supplemental nitrogen is applied. When woody yard waste, sawdust, bark, or other high carbon materials are incorporated into soil, supplemental nitrogen must be applied to prevent immobilization and nitrogen deficiency for plants (Table 3-4).

Sources of Soil Organic Matter

The fresh organic matter in most soils consists of plant and animal residues in various states of decomposition. Residues with the lowest carbon:nitrogen ratios decompose the most rapidly. Examples of low carbon:nitrogen residues that are often applied to soil are grass clippings, tree leaves, kitchen wastes, and animal manure. Some composts often have low C:N ratios but—having already gone through a process of decomposition while composting—are relatively stable and do not break down as quickly. Animal manure is a traditional source of plant nutrients. However, manure may contain high levels of salts, which can have deleterious effects on soil conditions and plant growth if applied at excessive rates or in poorly drained soils. Some manures and composts may have a high carbon:nitrogen ratio and temporarily immobilize nitrogen after addition to soil. Detailed information about organic matter sources for ornamental plants and turf can be found in Chapter 4.

Organic Soils

Organic soils are classified as those that contain more than 15 to 20 percent organic matter by weight, depending upon the soil texture. Organic soils develop when conditions are such that normal decomposition of soil organic matter is extremely slow. These conditions occur in river deltas, shallow lakes, swampy areas, and peat bogs. Organic soils appear almost black when moist. They are quite lightweight when dry and often have bulk densities of less than 0.5 grams per cubic centimeter compared to mineral soils that have bulk densities of 1.2 to 1.4 grams per cubic centimeter. Organic soils can hold large amounts of water—up to four times their weight. Organic soils can be quite productive if they are managed properly. Judicious use of nutrients and amendments is usually necessary to achieve optimum production. Nitrogen fertilization may not be needed initially, but after a few years of plant production it is usually necessary to apply nitrogen to sustain good plant growth. Phosphorus and potassium are usually low in organic soils and must be applied for optimum growth. Deficiencies of micronutrients, such as zinc, copper, and manganese, are more common in organic soils than in mineral soils.

The term organic here should not be confused with the same adjective that is used to describe a method of plant production that does not rely on inputs that are synthetic or from a petrochemical source. For information on "certified" organic soils or "organic" production, refer to Chapter 8.

SOIL ORGANISMS

In addition to their role in soil-forming processes, soil organisms make an important contribution to plant growth through their effects on nutrient cycling in soil. The most important groups of soil microorganisms are bacteria, fungi, and protozoa. In very wet soils, algae are also important. *Actinomycetes*—a class of microbes related to bacteria that forms fungal-like structures—are very common in soils and include both plant pathogenic and beneficial species. The number of individual microbes and microbial species in soil is quite astonishing. A teaspoon of soil typically contains from 100 million to 1 billion individual bacteria. Soil microbes are important in the decomposition of organic materials, the subsequent retention and release of nutrients, fixation of atmospheric nitrogen, and creation of soil aggregates.

Soil bacteria are of special interest because of their numerous varied activities. Heterotrophic bacteria, a large group of microorganisms, use organic carbon as a carbon source and are very active in decomposing plant and animal residues, releasing plant-available nutrients, and making humus. Autotrophic bacteria use carbon dioxide as a carbon source and obtain energy from the oxidation of materials such as ammonium, sulfur, and iron. Autotrophic bacteria are responsible for *nitrification*, the conversion of ammonium to nitrate, in the soil. The nitrification process is vital in providing nitrogen for plant growth.

In most soils, the use of supplemental microbial additives is not helpful for plant growth due to the high population and diversity of the existing microbial pool. There are some conditions where microbial additives may be beneficial for plant growth, such as in recently sterilized soils, sterile soilless media, or where additional nitrogen-fixing bacteria or mycorrhizal fungi are needed.

Nitrogen-fixing bacteria play an important role in the growth of higher plants. They are capable of converting atmospheric nitrogen into reduced nitrogen that is available to the plant in which the bacteria are living. Nodule bacteria (*Rhizobium* spp.) live symbiotically in the roots of a legume, derive their energy from the host plant, and fix nitrogen from the soil atmosphere. The amount of nitrogen that is fixed depends on many factors, such as the species of legume, the strain of bacteria infecting the roots, the overall plant health, and growing conditions.

SOIL MANAGEMENT

The importance of proper soil management and conservation for protecting natural resources is clear. Tillage practices can be an important tool for soil management. Some of the major reasons for tillage are preparing seedbeds, controlling weeds, incorporating plant residues and fertilizers, breaking soil crusts and hardpans, and improving soil characteristics for irrigation and erosion control. Inappropriate and ill-timed tillage can cause soil compaction, increase erosion, and reduce soil quality and productivity.

Sod production requires premier soil management. When properly performed, harvesting sod from the field removes a minimal amount of mineral soil. Sod production facilities intentionally encourage root growth in the upper layer of organic matter and avoid harvesting the mineral layer of soil. This is accomplished by first properly preparing the soil before planting, proper fertilization and irrigation practices to encourage shallow root growth, and then rolling the turf prior to cutting to provide a uniform cutting surface. Minimizing soil removal has the advantage of preserving the field production site, lowering turf roll weight, lessening transportation cost, and stimulating root recovery after laying the sod.

Soil conservation is an important management practice that deserves careful attention. It has been estimated that the cost due to annual soil erosion loss in the United States approaches $40 billion. Soil conservation practices have reduced the rate of topsoil loss from an average of 4.1 tons of soil lost per acre per year in 1982 to 3.1 tons in 1992. Terracing, rip-rap, wattles, dust control, hydroseeding, and

FIGURE 3-9. Profile of turf showing the layers of thatch and soil.

FIGURE 3-10. Careful soil and plant management is important for today and tomorrow.

ground cover are some of the important soil conservation practices that should be utilized to protect this valuable resource.

Proper residue utilization is key to good soil management. Plant residues returned to the soil help maintain soil productivity through maintenance of soil organic matter. Plant residues are one of the most valuable resources growers can use to minimize wind and water erosion.

Soil management is one component of a total landscape management system. Best Management Practices (BMPs) are those fertilizer, pesticide, and water management practices that lead to increased fertilizer efficiency, minimize the loss of inputs, and maintain or increase quality. The adoption and utilization of BMPs will result in both the greatest economic return to growers and the maximum protection of the environment. These are discussed in greater detail in Chapter 12.

Although soils in the US have historically been subjected to abuse and erosion that could have been prevented, much progress has been made in protecting this vital resource. We should exert every effort to manage them properly. *Manage* is defined as "handle with a degree of skill" or "treat with care," therefore good soil management must reflect these definitions. Proper management implies using the best available knowledge, technologies, materials, and equipment for growing plants. Through wise management, we can produce aesthetically pleasing plants, improve soils, and enhance the environment, thus providing a priceless legacy for future generations.

CONSTRUCTED SOILS

In many situations, people can use the naturally occurring soil as it exists with only minor modification. Typical soil modifications are applying fertilizers to supply missing nutrients, adding amendments to adjust soil pH, or applying composts to increase soil organic matter. In other situations, however, the intended use of the site requires a major modification of the soil.

Soils and Soil Substrates for Athletic Use

Most natural soils are not suitable for intensive uses such as in soccer and football playing fields and golf course greens and tee boxes. Intensively used fields often have inadequate drainage and the fields become excessively muddy after irrigation or rainfall.

The construction of natural grass athletic fields and other areas is a complex topic that requires site-specific information. The information in this section is designed to provide a brief overview of the major issues to be considered. Consult with local experts before embarking on a major soil construction project. Construction of golf greens should adhere to the United States Golf Association (USGA) specifications. There are a number of approaches to constructing successful sand-based athletic fields. These have been developed through formal research as well as informal "trial and error."

Good drainage is the crucial element in constructed soil fields. Many soils do not have adequate internal drainage and will benefit from additional subsurface drainage. A subsurface drainage system should be installed with lateral lines no more than 15 feet apart. The drains should be placed 16 to 24 inches below the soil surface in trenches. The trenches should be back-filled with pea gravel. The USGA requires a four-inch layer of pea gravel on the sub-grade, but this may be omitted for athletic fields. Coarse base sand is then placed on the gravel or sub-grade to promote water movement to the tile drains. The depth of this layer varies from 4 to 12 inches. The top layer of the constructed soil, the root zone mixture, is primarily or exclusively fine sand depending on the intended use of the field. The addition of small amounts of organic matter to the root zone mixture can increase desirable soil physical properties and is usually required for golf greens. Vigorously growing turf on an athletic field will contribute to the organic matter in the surface root zone and additions of supplemental organic matter may not be necessary.

Proper fertilization of sand-based fields is critical because sands are inherently infertile. Fertilization practices will depend on the turf species grown, growing conditions at the site, climate and weather patterns, and management objectives.

Proper management of constructed soils is essential for them to remain useful for their intended purpose. More information on fertilization is found in Chapter 12. However, management practices such as turf mowing height and frequency, irrigation, core aeration, vertical mowing, sand topdressing, and over-seeding must be done appropriately for each specific field. Consult with local authorities for the recommended timing of these practices in your area.

Soils for Shrubs and Trees for Transplanting

Trees and shrubs grown in the field for commercial production have specific soil requirements. For example, a loose soil, such as a sand or sandy loam, is preferred for bare-root operations so the soil can be easily removed from the roots with little root damage. These coarse soils require more frequent irrigation and retain nutrients poorly, so more attention must be paid to nutrient management. Ball and burlap operations, where perennials are field grown and dug before transporting to the planting site, often use a finer-texture soil—one with more clay and organic matter—so the ball will remain intact during the harvesting and wrapping process.

SOILS AND SUBSTRATES FOR HORTICULTURAL USES

In contrast to agricultural soils, which are used with only minimal modification, soils used for horticultural applications may be modified extensively to achieve specific physical and chemical characteristics. In many situations, there is no mineral constituent present and the material is referred to as soilless media, or more commonly, substrate. Regardless of the degree of modification, the soil or substrate must provide a suitable environment for plant growth and development. The soil or substrate must supply nutrients, store and release water to plants, permit exchange of gases between the substrate and the atmosphere, and provide mechanical support for the plant. The balance between water holding capacity and drainage is of particular importance for potting mixture and rooting substrates. The substrate must be able to hold and release water to the plants without becoming excessively wet, which can lead to root and crown diseases.

SUPPLEMENTARY READING

Biondo, R.J., and J.S. Lee. 1997. *Introduction to Plant and Soil Science and Technology.* Interstate Publishers, Inc., Danville, IL.

Brady, N.C. 1996. *The Nature and Properties of Soils,* 14th ed. Prentice Hall, Upper Saddle River, NJ.

Buol, S.W., F.D. Hole, and R.J. McCracken. 1997. *Soil Genesis and Classification,* 4th ed. Iowa State University Press, Ames, IA.

Hamrick, D., ed. 2003. *Ball Rebook, Vol. 2: Crop Production,* 17th ed. Ball Publishing, Chicago, IL.

Handreck, K., and N. Black. 2005. *Growing Media for Ornamental Plants and Turf,* 3rd ed. University of New South Wales Press, Ltd., Sydney, Australia.

Hudson, N. 1995. *Soil Conservation.* 3rd ed. Cornell University Press, Ithaca, NY.

Miller, R.W., and D. Gardiner. 2008. *Soils in Our Environment,* 11th ed. Prentice Hall, Upper Saddle River, NJ.

Nelson, P.V. 2008. *Greenhouse Operation and Management,* 7th ed. Prentice Hall, Upper Saddle River, NJ.

White, R. 2005. *Principles and Practice of Soil Science: The Soil as a Natural Resource,* 3rd ed. Blackwell Science, Oxford, UK.

CHAPTER 4

Soilless Media

Native soil is usually a poor growing medium for containerized plants because it lacks optimum aeration and makes the handling of containers difficult because they contain heavy moist soil. Therefore, media used for growing plants in containers is made up of a combination of lightweight, porous substrates and little, if any, soil. The diverse mixes made from a wide range of substrates are known as *soilless media*. Most greenhouses and container nurseries now use soilless media (Figure 4-1).

Soilless media provides adequate aeration while simultaneously supplying other needed physical properties such as water-holding capacity and added weight for container stability. Soilless media are created with organic and inorganic substrates. These

FIGURE 4-1. Soilless media is used in nearly all nurseries and greenhouses.

substrates may include coconut coir, rice hulls, perlite, vermiculite, rock wool, recycled materials, calcined clays, and wood residues. The availability of some products such as peat, redwood, and other barks is declining. The choice of substrates is dictated by specific physical and chemical characteristics, availability, and price. Locally available substrates are less expensive than products that must be shipped long distances.

PROPERTIES OF SOILLESS MEDIA

The primary role of any medium is to provide adequate water, oxygen, and nutrients. There are several physical and chemical properties to consider when making a soilless mix. In addition, a medium should not contain pathogens, toxic elements, excess salinity, or toxic biological compounds.

Physical Properties

Porosity is the percentage of the medium volume not filled with solids. Porosity is determined by the range and distribution of particles in the medium. After irrigation, a portion of these pores are filled with water while the remainder of the pores are filled with air. Soilless media porosity ranges from 25 to 90 percent.

Air-filled porosity is the volume percentage of the medium occupied by air. This parameter is influenced by the porosity of the medium and the container size and shape. A short container will have less air-filled porosity than a tall container filled with the same type of medium. This is because gravitational forces in a tall container will cause a smaller film of water to remain at the bottom of the container than in a short container.

Some landscape plants with high aeration requirements include the plant families of Ericaceae (blueberry, azalea, laurel, rhododendron, arbutus, manzanita, etc.), Proteaceae (protea), Thymelaeaceae (daphne, edgeworthia), and Theaceae (camellia, franklinia, stewartia). Additionally, orchids require media with a very high air-filled porosity. While aeration is essential for healthy root growth, media with excess aeration will not have sufficient water-holding capacity. Air-filled porosity in media can range from 10 to 50 percent.

Water-holding capacity is the volume percentage of water retained in a container after it has been fully irrigated and allowed to drain. This is also referred to as *container capacity*. Media with high water holding capacities require less frequent irrigation than media with low water holding capacities. However, if the water holding capacity is too high, adequate aeration could be compromised. Water holding capacity for typical media ranges from 20 to 70 percent.

Soilless media can be characterized as having either hydrophilic or hydrophobic properties. A dry medium that readily absorbs water is termed *hydrophilic*, which means "water loving." Coir (coconut husks) is a hydrophilic substrate. The opposite of hydrophilic is *hydrophobic*, which means "water hating." Peat is a hydrophobic substrate. Once dry, peat is very difficult to re-wet.

Bulk density is the weight of medium per volume. Physiologically, bulk density does not affect plant growth. Bulk density is a physical property that impacts production processes. If the bulk density is too low, the container is subject to tipping over, especially with taller plants. If the bulk density is too high, the cost of shipping the heavier containers increases. Bulk densities of soilless media range from 0.15 to 1.5 g/ml compared to soil, which ranges from 1.2 to 1.6 g/ml.

The *flowability* of a medium is the ability of the medium to flow along conveyor belts. Substrates that have poor flowability are shredded coconut husks, shredded redwood, and fibrous peats. Like bulk density, flowability only affects the mechanical aspects of production processes, but has no impact on plant health.

While not directly influencing media components, the *temperature* of the medium will impact the health of the root system. Usually roots do not develop the same level of cold hardiness as the shoots. Therefore, containers may need to be insulated from the cold during the winter. During the summer, the medium can become very hot from exposure of container sides to direct sun. Media temperatures of 120°F have been measured when outside air temperatures were 85°F. Heat stress is especially a problem on the southwest sides of containers. To minimize heat stress, use lighter colored containers that will not absorb as much heat from the sun or protect the southwest sides of containers from direct sun.

TABLE 4-1
Physical Properties of Two Container Media at Container Capacity[1]

Medium Mixes	Percent by Volume				
	Total Porosity	Air Porosity	Water Content	Dry Density	Wet Density
				(g/ml)	*(lb/cu ft)*
1 Sand + 2 Peat	74	10	64	0.64	80
1 Perlite + 1 Peat	94	20	74	0.13	54

[1]Depth of medium = 5 in

The effects of media components on media physical properties are demonstrated in the example shown in Table 4-1. Inclusion of sand in the medium decreases its porosity and increases its density, compared to a medium comprised of lightweight perlite.

Chemical Properties

Electrical conductivity (EC) is a measure of the total soluble salts that are extractable from the medium. Most containerized plants can tolerate soluble salts with an EC between 2.0 and 3.0 dS/m. Lists of salinity tolerances of many ornamental plants can be found in Chapter 5. The amount of fertilizer in the medium has a major impact on the EC. However, it is the nonessential salts, such as sodium, that can have a detrimental effect on root growth. Some coir products soaked in brackish waters or manures not properly processed will have a high EC that can be detrimental to plant growth.

Cation exchange capacity (CEC) is the capacity of the medium to adsorb positively charged elements such as calcium (Ca^{2+}), magnesium (Mg^{2+}), ammonium (NH_4^+), and potassium (K^+) onto the negatively-charged surfaces of the particles. The CEC of media is measured on charge per unit weight basis as cmol(+)kg^{-1} or on a charge per unit volume basis as cmol(+)L^{-1} (previously, milliequivalents per liter). Most organic matter used in media is not decomposed enough to have a high CEC. The CEC of a medium is not usually an important consideration when choosing a substrate since containerized plant production processes rely on nutrients added in granular fertilizers or through irrigation systems.

The medium *pH* is a measure of hydrogen ion (H^+) concentration in the medium. The pH can impact plant growth by: 1) affecting the integrity of the growing root; 2) affecting the availability and transport of nutrients to the root system; and 3) changing populations of beneficial and pathogenic organisms that can enhance or interfere with root growth. A very low (<4.0) or high (>8.0) pH, depending on the plant species, can directly damage the cells of growing roots.

Regarding plant nutrition, a lower pH tends to increase the solubility of certain micronutrients such as iron (Fe), copper (Cu), zinc (Zn), and manganese (Mn). Soilless media have a maximum availability of nutrients at a lower pH, between 5.0 and 6.0 (Figure 4-2),

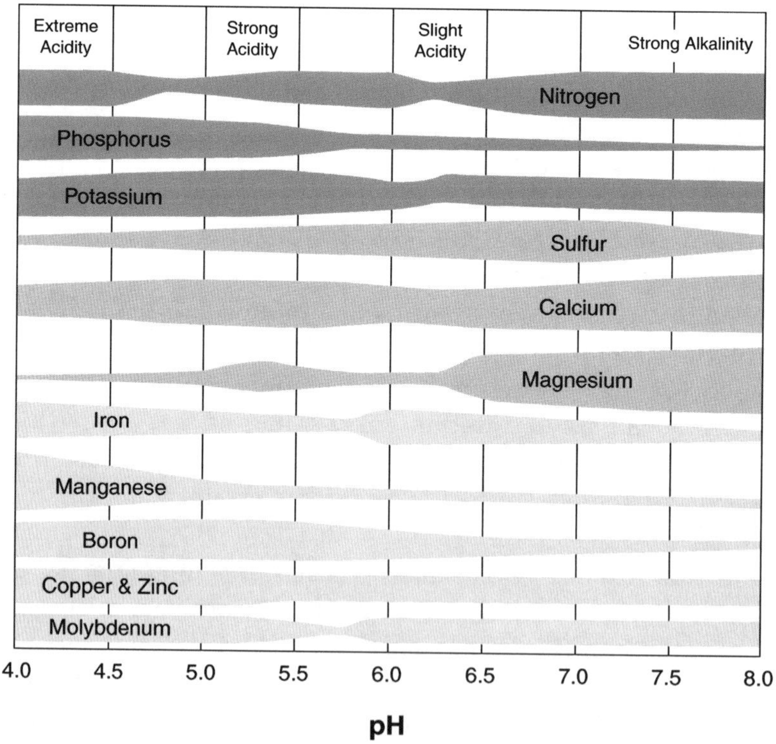

FIGURE 4-2. Relative mineral nutrient availability in soilless media as affected by pH. *Source:* Handreck, K.A., and N.D. Black. 1999. *Growing Media for Ornamental Plants and Turf.* UNSW Press, Sydney, Australia.

compared to mineral soils where optimum nutrient availability ranges between 6.5 and 7.5 (see Chapter 3, Figure 3-7). The difference in optimum nutrient availability between mineral soils and soilless media lies in the effect pH has on the cation exchange capacity (CEC) of organic matter.

Liming growing media, which raises media pH, is more complex than liming soils. See Chapter 6 for liming soils. Many labs and soilless media manufacturers can determine lime requirements of a soilless media for their customers. An individual can determine lime required to raise media pH by performing an incubation study. One procedure to determine media liming requirements is as follows:

- Divide 2 cubic feet of potting mix containing fertilizers into four ½ gallon portions
- To each portion add and thoroughly mix 0.5, 1, 2 and 3 ounces of liming material
- Add water, preferably irrigation water, until each portion is slightly wetter than normal
- Store each portion in a clean, re-sealable plastic bag at room temperature out of direct sunlight
- Remix the samples every 3 to 5 days
- Remove sub-samples after two weeks of incubation and determine the pH
- Plot the resulting pH against the amounts of lime added
- Use the chart to determine the amount of lime required
- Multiply ounces of lime per 0.5 cubic feet by 3.375 to determine pounds of lime required per cubic yard

Biological Properties

There are *beneficial microbes*, such as *mycorrhizae* or *trichoderma*, which can colonize a root system and aid healthy root development as well as nutrient uptake. However, it is difficult to maintain populations of these organisms under typical growing conditions. Most nursery and greenhouse containerized plant production facilities apply high rates of fertilizers that prevent the establishment and survival of beneficial microorganisms.

Certain organic chemicals present in some substrates may impact plant growth. *Allelopathic compounds*, which are those chemicals produced by some plants to limit the growth of other plants, may be present in some media components and prove deleterious to plant health if these compounds are available at harmful concentrations.

Most substrates will need to be sterilized before use to kill pathogens, weed seeds and propagules, and insects. Proper composting temperatures of 140°F for at least 24 hours are required to kill most unwanted organisms.

WATER RELATIONS OF SOILLESS MEDIA

Water retention and aeration varies significantly between soilless media and soil once they are placed in a container. Drainage from the bottom of a container is governed principally by factors affecting hydraulic conductivity. Water pressure due to gravitational forces at the bottom of a container also plays a role. Water pressure increases by 0.433 pounds per square inch for each foot in height. In practice, reducing the height of a column of media reduces drainage and increases water retention because reducing the height of the wetted media reduces pressure at the base of the column.

Container media must drain well. Every living cell in the plant, including roots, requires oxygen. Soilless media should provide adequate aeration balanced with adequate moisture for plant use without having to irrigate frequently. Perlite and vermiculite create large pores that accentuate drainage. On the other hand peat moss, once wetted, is particularly good at increasing moisture retention so sufficient water is available to the plant roots. The desired degree of aeration and container capacity will dictate the balance of components used to create a specific media mix. The aeration requirements of some ornamental plants are listed in Table 4-2.

The amount of water in a container will not be equal from the top to the bottom of the container. The amount of water at the surface of a container may be very little as compared to saturated media that may form towards the bottom of the container. Drainage out of the bottom of the container, assuming drainage holes are not

TABLE 4-2
Approximate Root Aeration Requirements of Selected Ornamentals Expressed as Percent Air Space

Very High 20%	High 10–20%	Intermediate 5–10%	Low 2–5%
Azalea	African violet	Camelia	Carnation
Fern	Begonia	Chrysanthemum	Conifers
Orchid, epiphytes	Daphne	Gladiolus	Geranium
	Foliage plants	Hydrangea	Ivy
	Gardenia	Lily	Palm
	Gloxinia	Poinsettia	Rose
	Heather		Stocks
	Orchid, terrestrial		Strelitzia
	Podocarpus		Turf
	Rhododendron		
	Snapdragon		

plugged, will only occur if the media at the base of the container is saturated or if the wetted zone is open to a source of positive pressure from above.

Irrigation practices must be monitored closely and one must prepare media that allows for drainage. Vermiculite, perlite, and peat moss in media mixes create large pore spaces to increase drainage, are light in weight, and increase container capacity due to the large amount of surface area. Many laboratories have procedures that can test soilless media for porosity, water holding capacity, and aeration.

SOILLESS MEDIA COMPONENTS

There are many different types of organic and inorganic components used to create a soilless media (Figure 4-3). The advantages and disadvantages of the different substrate types are described. In

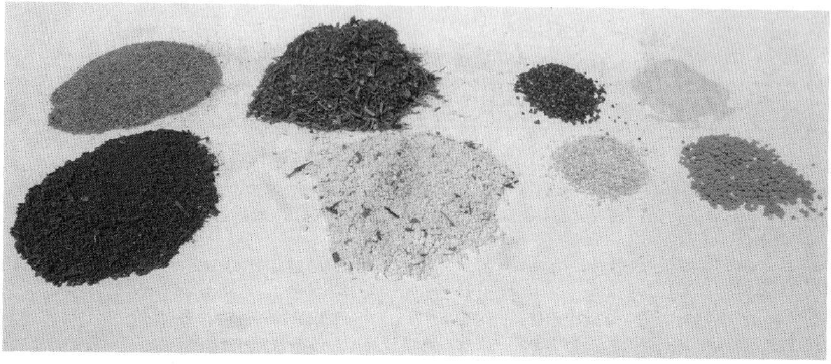

FIGURE 4-3. Soilless media is comprised of numerous components.

many cases, substrates are blended together in different ratios to optimize the physical and chemical characteristics of the medium. The diverse needs of the many kinds of plants grown in nurseries and greenhouses require different soilless media mixes. Not only are the physical and chemical properties of the individual media components important, but many of the components may also contain plant nutrients as well as elements that might interfere with balanced nutrient uptake. The nutrient content of the final mix will be important when a fertilizer program is designed to meet plant nutrient needs.

Organic Soilless Media Components

The availability of different types of organic components depends on several factors. Soilless media substrates that are available close to the end user are less expensive than those that must be shipped from a greater distance. Peat moss is becoming more expensive as supplies from natural resources decrease and demand increases. Other products, such as municipal compost, are becoming more available as disposal fees increase and available landfill space decreases. All organic products should be properly sanitized so that all pathogens and weed propagules are killed. Typical ranges of properties of organic soilless media components can be found in Table 4-3.

Peat

Peat is composed of plant residues that have accumulated and undergone varying amounts of decomposition in water or in excessively

TABLE 4-3

Typical Properties for Organic Soil Amendments

Organic Amendment	Moisture Retention	pH	Organic Matter	Ash
	(% total volume)		(%)	(%)
Sphagnum peat	60–70	3.2–4.5	95–99	1–5
Hypnum peat	55–65	4.4–6.7	70–85	15–30
Reed and sedge peat	50–60	4.5–7.0	85–95	5–15
Woody peat	30–40	3.6–5.5	75–90	10–25
Sawdust	30–40	3.8–8.0	95–99	1–5
Ground bark	30–40	4.0–8.0	90–95	5–10
Compost	20–30	4.0–8.0	80–85	15–20
Leaf mold	20–30	4.0–8.0	50–75	25–50
Coir	60–70	5.5–6.5	94–98	5–10

wet areas, such as swamps and bogs. Peat exhibits a wide range of water-holding capacities, organic content, pH, and longevity. Peat is hydrophobic; once it becomes dry, peat is difficult to re-wet. A surfactant or wetting agent may be required to re-wet the medium. Peats are classified into different types according to the kinds of plants from which they were derived, the length of time they have decomposed, and the geographic area and depth of bog from which they came. One or all of the following peat layers may be present in bogs or swamps.

Moss, the topmost layer of a bog, is not a true peat and is generally scraped off and discarded before harvesting peat from a bog.

Moss peat is partially decomposed peat located directly under the moss layer in a bog and is very low in plant nutrients. Moss peat has a high water-holding capacity—equal to 15 to 30 times its own weight—and is lightweight, porous, and difficult to mix into soils. It is light brown to tan in color and decomposes at a moderate rate because of its high cellulose content. Moss peat is used in hanging baskets due to its high water-holding capacity.

Sphagnum peat, also known as *peat moss,* has the highest commercial value in the turf, greenhouse, nursery, and landscape industries. This layer is partially decomposed and lies below moss peat in a bog. Sphagnum peat is generally more fibrous in nature and is blonde in color. Sphagnum peat can be purchased in several grades from fine to coarse. Fine peat is suitable for media blends used in plug production but holds too much water when used in larger containers. Peat may harbor soil pathogens, such as *pythium* and *fusarium*. Steam sterilization of peat by manufacturers is recommended.

Sedge peat accumulates in swamps, along the edges of bogs, and beneath the sphagnum peat in bogs and has undergone longer decomposition than sphagnum peat. Reddish to dark brown in color, this layer in a bog may be more fibrous than sphagnum peat. Sedge peat does not have as high a moisture-holding capacity as the upper layers in a bog. Sedge peat is less acidic than sphagnum peat.

Hypnium peat is the deepest and most decomposed layer in a bog and is not usually used in the horticulture industry.

Forestry By-Products

Tree residues such as sawdust and bark come in various grades or sizes and are widely used in preparing container soil mixes. Smaller grades of bark provide good water-holding capacity, whereas larger grades of bark residues possess poor to fair water- and nutrient-holding capacities. In contrast, sawdust and fine residues may exhibit excessive water-holding capability. The rate of decomposition varies with the size and kind of wood residue used. Residues from soft pines may decompose after several months and should be used only when fully composted. Residues from fir, redwood, cedar, and cypress may last up to five years. Tree residues containing wood will require more nitrogen than most bark products, as bark usually breaks down slower. Ground fir bark has proven to be a valuable component in container mixes. The round shape of the bark provides good water and oxygen diffusion rates and resists compaction. Fir bark has low nutrient-holding and fair water-holding capacity.

Coir or *coco peat,* chopped and ground coconut husks, is a by-product of the coconut industry. Coir consists of lignified fibers and pith. The fibers of the coir provide drainage and the pith is excellent for water retention. Fibers are often removed and used to make rugs, rope, and other products. If too much fiber has been removed during processing, the high percentage of pith remaining in the coir makes the media too wet when used in containerized plant production. Therefore, it is important to see that enough fibers are present in coir products to maintain adequate drainage. Coir chips do not have this problem. It is only the fine coir products that have too much absorptive capacity.

Early users of coco peat encountered problems with high salt levels because the husks were left in brackish waters until processed. These problems have been largely resolved, but check the electrical conductivity and request a chemical analysis from the supplier. Coir is gaining popularity because of improved quality assurance, cost, and favorable physical characteristics that include good moisture-holding capacity, cation exchange capacity, porosity, a slightly acidic pH, and hydrophilic properties. Coir comes as compressed bricks that require hydration to decompress. It may also come as bales of different grades of coir chips, which are often used in hydroponic systems or for orchid production. Some coir contains naturally occurring *trichoderma*, a beneficial fungus, which has been shown to suppress common root diseases. Peat-coir blends are also available, which have excellent media characteristics since the normally hydrophobic nature of the peat is eliminated with the hydrophilic coir and the low pH of the peat is neutralized with the higher pH coir.

Agricultural and Municipal Wastes

Rice hulls are the tough outer coatings of rice grains. The hulls decay very slowly in media, lasting up to 10 years, making them an excellent substrate for perennial container plants that are in production for several years. Rice hulls, however, have neither good water- nor nutrient-holding capacities. Rice hulls are relatively inexpensive but these physical characteristics preclude them from comprising more than 30 percent of most media blends.

Composted landscape waste is becoming more available. While inexpensive, it is important that these products are:

- Not contaminated with damaging concentrations of herbicide residues
- Properly composted so that shrinkage of the medium does not occur
- Not high in electrical conductivity

- Free of viable weed propagules (seeds and stems) and pathogens
- Uniform in particle size distribution
- Readily available on a regular basis

Agricultural wastes are sometimes included in soilless media mixes. Examples include different types of nut shells, sugar cane residue, and straw. Depending on the agricultural commodities close to the end-user, other agricultural wastes may be utilized. As long as these products are properly composted, they may be used with other substrates in media formulations to develop a medium with suitable physical and chemical traits for the plants being grown.

Manure or animal wastes may be used in media mixes but they can be high in salts and urea. If not properly composted and leached, excess salinity can chemically "burn" roots. Manure is an excellent source of nitrogen but much of the nitrogen is readily available and easily leached.

Sludges are the residues from sewage treatment facilities. These products may be a good source of nitrogen, but the fineness of the particles may cause drainage problems in container mixes. In some cases, heavy metals may also be of concern. Despite the low cost of sludge, it is not normally recommended as a soilless media substrate.

Inorganic Soilless Media Components

Inorganic substrates are useful in media since these products rarely decompose. Therefore, coarse inorganic substrates are often used to enhance and maintain the porosity of media, especially for plants requiring greater than a one year production cycle. Inorganic substrates are generally not a source of plant nutrients, but may aid in nutrient retention and root health by improving porosity, aeration, and drainage.

Three important questions to ask when choosing an inorganic substrate are:

- Does the substrate float or sink in water?
- How abrasive is the product on equipment?
- How physically stable is the substrate in planting processes?

Calcined clay is a granular substrate that resists compaction, does not float, and is moderately abrasive. Similar to kitty litter, calcined clay adds weight to containers that will hold tall plants that might fall over with a lighter mix. Calcined clay also increases porosity and water-holding capacity of media. Even though it has a high water-holding capacity, the water is tightly held to the clay particles limiting the water availability to plant roots. Unlike regular clay, calcined clay has a poor cation exchange capacity.

Perlite is a stable, white, bead-like particle that is produced from heat-treated volcanic rock. It is a low bulk density substrate used to provide porosity and drainage to media. Perlite is often used in combination with peat moss as a medium for ornamental plant propagation. One negative physical characteristic of perlite is that it floats in water, which can be a problem in certain production scenarios. Perlite is moderately abrasive.

Pumice is a ground, lightweight, porous volcanic rock with a sponge-like appearance. Although expensive, pumice improves water infiltration and porosity of media. Water- and nutrient-holding properties of pumice are fair to poor. Some pumice products may have excessively fine pores that retain water too tightly causing limited water availability to plants. Pumice is abrasive.

Rockwool is made from melted basaltic rock that is spun into thread-type blocks. The blocks or sheets of rockwool are then cut into cubes or formed into plug trays. Unprocessed rockwool is normally hydrophobic. Therefore, wetting agents are added to rockwool when it is produced for horticultural purposes. If rockwool contains impurities, such as slag, residues may be present that could be toxic to plants. The benefits of rockwool are that it is uniform, porous, stable, and has a high water-holding capacity. One drawback is that rockwool retains excessive amounts of water and may impede drainage. Rockwool is moderately abrasive and a skin irritant.

Sand has low water- and nutrient-holding capacities. It is used in media when additional weight is needed for container stability or to increase drainage. However, the proper quantity of sand required in soilless media depends on the size of the sand particles. Finer sands impede drainage more than coarse sands. Sand is abrasive.

Vermiculite is made by super-heating a flaky, micaceous mineral. The internal moisture of the mica turns to steam and expands the flakes to many times their original size. The grade or size of the vermiculite is determined by the size of the ore that is expanded. The resultant fluffy, lightweight product can hold several times its own weight in water. Even when thoroughly wet, there is still sufficient air for plant roots to grow. These properties make vermiculite ideal for topping media when plants are grown from seed. The vermiculite acts like a vapor barrier, slowly and evenly giving up its moisture to help maintain high humidity near the seed. Vermiculite is permissive to light and acts as an insulator. Vermiculite has some nutrient-holding capabilities but has a low cation exchange capacity. It lasts indefinitely if properly handled. Vermiculite is included in vermiculite-peat products and potting soil mixes, and is used in specialized greenhouse propagation. Vermiculite is not abrasive.

Expanded shales, clays or *slates* are made by super-heating flaky minerals in a rotary kiln to create lightweight, porous materials useful for improving aeration and water-holding capacity. These products are very stable and mildly abrasive. Their use is increasing in special projects where lightweight artificial "soil" is required, such as in planting boxes and rooftop gardens. These components are used as soil amendments in baseball diamonds and golf courses.

New inorganic substrates are constantly being tested. Some of the precautions with inorganic substrates are:

- Contamination by heavy metals or synthetic residues toxic to plants
- Substrate physical stability in media mixing
- Worker safety
- Potential risks of environmental contamination

NUTRIENT REQUIREMENTS FOR SOILLESS MEDIA

Both organic and inorganic media components do not have high nutrient-supplying or nutrient-holding capacities. Therefore, soilless mixes require added nutrients. Most mixes will include a pre-mix formulation that contains fertilizers and amendments that balance the elements and chemistry of the soilless media (Figure 4-4). Many commercial nursery operations are on a constant fertilizing program, either through liquid feed, controlled release granular fertilizers, or both.

Pre-Mix Materials

Dolomitic limestone is commonly added to provide essential calcium and magnesium, as well as provide a pH buffering system for the medium. The natural conversion of ammoniacal fertilizers to nitrate can rapidly drop media pH to sub-optimal levels without the addition of limestone. Two or more particle sizes of limestone may be used to provide both a short-term and long-term buffering effect, with small and large particles, respectively. Application rates of limestone should be determined by the procedure outlined previously in this chapter, by an analytical lab, or by a media manufacturer.

Gypsum may be added to the medium as calcium and sulfate-sulfur nutrient sources where extra calcium is required but a liming effect is not desired. Gypsum is slowly soluble, releasing calcium and sulfate over time. Plants that will remain in containers for a year or more may require a larger particle sized gypsum for extended feeding. Rates of gypsum application can range from 1 to 3 pounds per cubic yard.

Slow or controlled release fertilizers may be added to the mix to provide a range of nutrients in between water-applied fertilizer feedings. Nitrogen is the macronutrient most needed by plants and also the most likely to be lost due to leaching. Therefore, either slowly-soluble nitrogen or coated nitrogen is added to media during the mixing process. Controlled release fertilizers containing a complete nutrition package are also available. Application rates for slow and controlled release fertilizers are typically included with the fertilizer label. For more information on slow and controlled release fertilizers, see Chapter 7.

Superphosphate (0-45-0) is commonly added to media. Superphosphate fertilizer supplies phosphorus, which is necessary to stimulate early root development. Superphosphate is useful for plant establishment as well as supplying calcium and sulfur. Application rates of superphosphate range from 1 to 2 pounds per cubic yard.

FIGURE 4-4. Commercial soilless media pre-mix operation.

Micronutrients, if not already included in the controlled release fertilizer formulation, may be added to the medium. The goal is to supply micronutrients that are either lacking in the media or to balance elements that are in excess in the mix. Many organic materials, such as pine and redwood bark, are high in iron and manganese and require a source of zinc. It is best to have a sample of the media analyzed by a laboratory before developing a pre-mix to make sure the right nutrients are added to create a balanced final mix. Most micronutrient products will have a recommended rate per cubic yard on the label.

Liquid Fertilization

Most plants grown in soilless media will require supplemental fertilization beyond the nutrients supplied by the media components and the pre-mix fertilizer. Application of fertilizers through irrigation water is known as *fertigation*. Fertigation entails injection of a fertilizer solution into the irrigation water before application to containers. Most systems have a batch tank containing the liquid fertilizer, a metering pump to precisely deliver a given rate of fertilizer, and an injection port in the main irrigation line to allow for the fertilizer to thoroughly mix with the water before irrigating plants.

Fertilizers for fertigation of soilless media are usually high quality, technical, solution, or greenhouse grade materials that are more highly refined than similar fertilizers used for general agriculture. Many greenhouse fertilizers come as soluble granules or crystals that readily dissolve in water with agitation. There are many formulations available for specific nursery and greenhouse needs. Some production houses may use two or three different formulations that are designed to provide different ratios of nutrients for different stages of plant growth.

The source of nitrogen is one important factor when deciding which fertilizer blend to use for soilless media. Many growers will avoid urea as a source of nitrogen for fertigation. This is because urea is considered to be an ammoniacal, and therefore, acidifying source of nitrogen. Urea is also non-polar and highly soluble and may leach through containers before it has a chance to be converted to ammonium-nitrogen. Strong root development with slow shoot growth is commonly the goal of greenhouse and nursery plant production. Ammoniacal sources of nitrogen tend to favor shoot growth over root growth, since the plant requires less energy to assimilate ammonium-nitrogen than nitrate-nitrogen. Ammonium-nitrogen may also build up in the media resulting in ammonium toxicity. For this reason, many nursery and greenhouse liquid fertilizer formulations are based on high proportions of nitrate-nitrogen derived from potassium nitrate, liquid ammonium nitrate, and calcium nitrate.

Formulations to meet soilless mix requirements should be obtained from commercial consultants or university horticultural advisors. Refer to the supplementary reading for more information.

SUPPLEMENTARY READING

Hamrick, D. (ed.) 2003. *Ball Red Book, Crop Production Volume 2, 17th ed.* Ball Publishing, Batavia, Illinois.

Handreck, K.A., and N.D. Black. 1999. *Growing Media for Ornamental Plants and Turf.* New South Wales University Press, Sydney, Australia.

Nelson, P.V. 2003. *Greenhouse Operation and Management, 6th ed.* Prentice Hall, Upper Saddle River, New Jersey.

Watson, J.R. 1967. Peat classifications. California Turfgrass Culture. 17(3): 21–24.

CHAPTER 5

Water and Plant Growth

Irrigation water is a crucial input growers control in plant production. Many plant functions rely on the unique characteristics of water. A major portion of the total plant fresh weight is water. Water keeps the plant turgid so it may better capture sunlight for photosynthesis, which combines water and carbon dioxide to form glucose. Water transports dissolved nutrients from the soil solution into the root, through the vascular system, and to all portions of the plant. The value of turf and ornamental plants is often based on appearance. The quality of water used in plant growth can have a major impact on the visual quality of the final product.

All water used for irrigation contains some dissolved salts. Water suitability depends on the kinds and amounts of salts present and the plants that are irrigated with that water. All dissolved salts have an effect on the properties of soils, growing media and, consequently, plant growth. The physical properties and behavior of soils and growing media are quite different. For more information on soils and soilless media, see Chapters 3 and 4, respectively.

A user of irrigation water should know the effects the following may have on soil or growing media:

- Salt content (salinity)
- Sodium status (sodicity)
- Alkalinity and pH
- Rate of water infiltration

- Toxic elements in plants, soil, and drainage water
- Plant nutrients in the water
- Soil nutrient status and groundwater quality
- Fouling or plugging of microirrigation systems

IRRIGATION

Irrigation is the replacement or supplementation of rainfall with water from another source in order to grow plants. Irrigation water is applied to soil or media in a number of ways to replenish water removed by evaporation, transpiration, and drainage from the root zone. The method of application depends on many factors such as the plants to be grown, depth and texture of the soil or media, container type, topography of the land, and the cost of both irrigation equipment and water. The amount of water used and how often it is applied are determined by water availability, the method of irrigation, plant needs, leaching requirements, the soil or growing medium water-holding capacity, and climatic considerations of the growing environment (i.e. greenhouse, lathhouse, nursery, or field

FIGURE 5-1. Irrigation water is applied to replace water removed from the soil.

conditions). Therefore, successful irrigation requires careful management of plants, soils or growing media, and water.

SOIL WATER BEHAVIOR

Pores between soil or media particles contain either air or water. In the turf and ornamental industry, the goal is to correctly balance air-filled pores with sufficient water to meet plant demand. Immediately following irrigation, the film of water is thick and not tightly held. Most of the air in voids between particles has been displaced by water. The amount of water held is called the *saturation percentage*.

Weakly held water in larger pores drains freely due to gravity. In turf and ornamental landscapes, the soil water content is at *field capacity* when free drainage ceases. Likewise, potted media has reached *container capacity* when the pot no longer freely drains following saturation. At field or container capacity, the film of water is thinner and more tightly held. Between 10 and 50 percent of the pores should be filled with air immediately after drainage has stopped.

Gravity is no longer a significant force in moving water through the soil profile when soil water contents reach field or container capacity. Roots of growing plants will remove most of the water by transpiration. But there comes a point when the remaining water is held so tightly by the soil or media that plants cannot extract it causing them to irreversibly wilt. This water content level is called the *permanent wilting point*. *Plant available water* is the amount stored between field or container capacity and permanent wilting point.

The water content of a saturated soil (i.e. its saturation percentage) is about twice the field capacity and about four times the permanent wilting point. This relationship among the saturation percentage, the field capacity, and the permanent wilting point generally holds consistent for all soils, from clay loams to sandy soils. Figure 5-2 schematically presents these relationships for soil. In well-drained soil or media, most water is held as a film around each particle. The thinner the film, the tighter water is held by the particle. As a result, increased energy is needed to remove water from the particles. The energy required to remove water from the surface of a particle is known as *suction* or negative pressure. Note

Soil Moisture

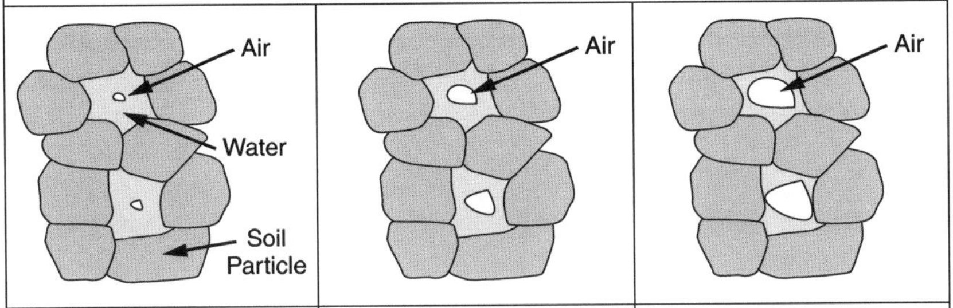

Saturation Percentage	Field Capacity	Permanent Wilting Point
1/10 Atmosphere	1/3 Atmosphere	15 Atmospheres
Near Saturation	1/2 Saturation Percentage	1/4 Saturation Percentage

FIGURE 5-2. Illustration of suction at and relationships between soil saturation percentage, field capacity, and permanent wilting point of a loam soil.

that soil suction values in Figure 5-2 represent a loam soil. Suction values for coarse textured soils are less than fine textured soils. For media, water content relationships depend on the ratio of materials used and their water retention properties.

Very moist soil has thick films of water surrounding particles and low suction. A drier soil has a thin film of water and high suction. For this reason, water will move from a wet soil with low suction to a drier soil with high suction, but the rate of such movement is slow.

Table 5-1 shows the water-holding capacity of different soils. Plants consume, on average, from 0.1 to 0.3 inches of water per day.

For media, the relationship between water content, container capacity, and permanent wilting point is harder to predict. Since most media are composed of a high percentage of organic material, it has a very low bulk density. How media is packed in a container, the components of the media, container geometry, and settling of the media over time have a large impact on the final bulk density (Table 5-2). Bulk density, in turn, has a large effect on the container capacity. Organic materials bind water films tighter

TABLE 5-1

Approximate Amounts of Available Water Held by Different Soils

Soil Texture	in Water/ft Soil	Max. Intake Rate (Bare Soil) in/hr
Sand	0.5–0.7	0.75
Fine sand	0.7–0.9	0.60
Loamy sand	0.7–1.1	0.50
Loamy fine sand	0.8–1.2	0.45
Sandy loam	0.8–1.4	0.40
Loam	1.0–1.8	0.35
Silt loam	1.2–1.8	0.30
Clay loam	1.3–2.1	0.25
Silty clay	1.4–2.5	0.20
Clay	1.4–2.4	0.15

TABLE 5-2

Effects of Container Filling and Packing Method on Media, Water, and Air Space

		Container	
Peat:Vermiculite (1:1)	6 in Standard	4 in Standard	48 Cell Flat
		Filled and brushed	
Available water	43%	51%	58%
Unavailable water	21%	21%	21%
Air space	23%	15%	9%
		Filled, tapped twice on bench	
Available water	44%	52%	52%
Unavailable water	26%	26%	26%
Air space	15%	9%	4%
		Filled, pressed and refilled	
Available water	45%	49%	52%
Unavailable water	30%	30%	30%
Air space	9%	4%	2%

Reed, D. W. 1996. *Water, Media, and Nutrition for Greenhouse Crops.* Ball Publishing, Batavia, IL.

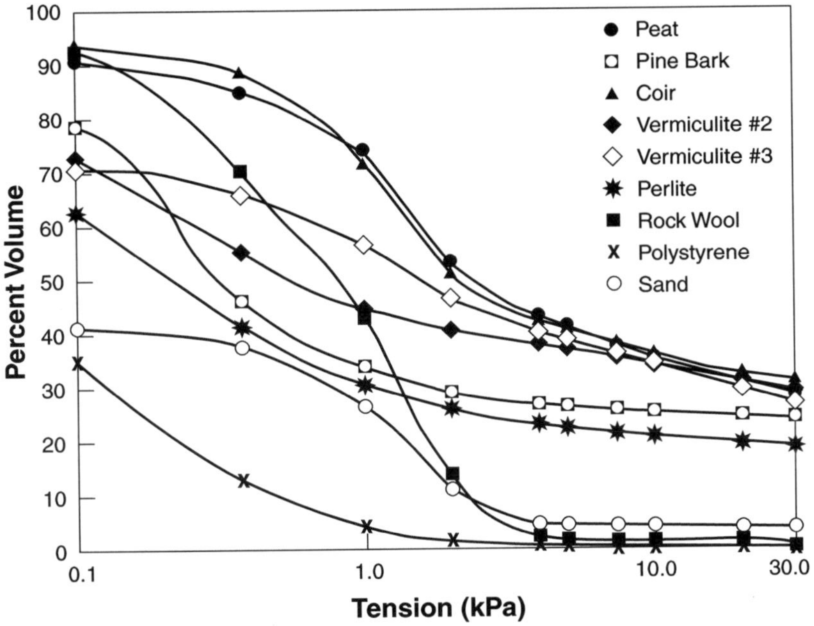

FIGURE 5-3. Moisture retention curves for soilless media components.

than mineral soils. This means media will typically reach permanent wilting point at higher water content than mineral soils (Figure 5-3).

WHEN TO IRRIGATE

Plant appearance is often used as a guide in determining when to irrigate. Symptoms such as slow growth, "bluish" color of leaves, and temporary afternoon wilting are signs of moisture stress. Plants should be irrigated before moisture stress symptoms are observed.

A soil tube or an auger can show depth of wetting and depletion of soil moisture. Experienced growers can determine the available moisture fairly accurately by feeling soil samples (Table 5-3). However, this technique is quite subjective and although it may work for plants grown under field or nursery bed conditions, it is not practical for container-grown plants. The most accurate moisture determination can be made by drying weighed samples of

TABLE 5-3
Soil, Moisture, Appearance, and Description Chart

	Feel or Appearance of Soil[2]			
Available Water[1]	**Sand**	**Sandy Loam**	**Loam/Slit Loam**	**Clay Loam/Clay**
Above field capacity	Free water appears when soil is bounced in hand.	Free water is released with kneading.	Free water can be squeezed out.	Puddles. Free water forms on surface.
100% (field capacity)	Upon squeezing, no free water appears in soil, but wet outline of ball is left on hand. (1.0)	Appears very dark. Upon squeezing, no free water appears on soil, but wet outline of ball is left on hand. Makes short ribbon. (1.5)	Appears very dark. Upon squeezing, no free water appears on soil, but wet outline of ball is left on hand. Will ribbon about 1 inch. (2.0)	Appears very dark. Upon squeezing, no free water appears on soil, but wet outline of ball is left on hand. Will ribbon about 2 inches. (2.5)
75–100%	Tends to stick together slightly. Sometimes forms a weak ball with pressure. (0.8-1.0)	Quite dark. Forms weak ball that breaks easily. Will not stick. (1.2 to 1.5)	Dark color. Forms a ball. Is very pliable. Slicks readily if high in clay. (1.5 to 2.0)	Dark color. Easily ribbons out between fingers. Has slick feeling. (1.9 to 2.5)

50–75%	Appears to be dry. Will not form a ball with pressure. (0.5-0.8)	Fairly dark. Tends to ball with pressure, but seldom holds together. (0.8 to 1.2)	Fairly dark. Forms a ball. Somewhat plastic. Will sometimes slick slightly with pressure. (1.0 to 1.5)	Fairly dark. Forms a ball. Ribbons out between thumb and forefinger. (1.2 to 1.9)
25–50%	Appears to be dry. Will not form a ball with pressure. (0.2-0.5)	Light colored. Appears to be dry. Will not form a ball. (0.4 to 0.8)	Light colored. Somewhat crumbly, but holds together with pressure. (0.5 to 1.0)	Slightly dark. Somewhat pliable. Will ball under pressure. (0.6 to 1.2)
0–25% (0% is permanent wilting)	Dry, loose, single-grained. Glows through fingers. (0-0.2)	Very slight color. Dry, loose. Flows through fingers. (0 to 0.4)	Slight color. Powdery, dry, sometimes slightly crusted, but easily broken down into powdery condition. (0 to 0.5)	Slight color. Hard, baked, cracked. Sometimes has loose crumbs on surface. (0 to 0.6)

[1]Available water is the difference between field capacity and permanent wilting point.

[2]Numbers in parentheses are available water contents expressed as inches of water per foot of soil depth.

moist soil in an oven to determine the percentage of moisture loss. A more practical method involves daily weighing of pots to determine losses or gains in container moisture. This method becomes less accurate as containers become root-bound.

SOIL OR MEDIA WATER MEASUREMENTS

Most moisture sensors indirectly measure soil moisture content by determining a correlated physical property. *Tensiometers* directly measure the amount of soil suction. Both direct and indirect methods have strengths and weaknesses. Properly placed and calibrated, soil moisture sensors can estimate the moisture content of soil and can help to schedule irrigations. A drawback, however, is that they are limited to moisture readings in predetermined locations and depths only. Moisture readings must be made at regular intervals, which can be done manually or with remote data loggers. Figure 5-4 illustrates the use of a tensiometer to determine soil moisture.

Examples of four commonly used soil moisture sensors are illustrated in Figure 5-5 (page 96).

A tensiometer consists of a porous ceramic cup that is placed in the soil or root zone. A rigid tube is connected to the cup and to a vacuum gauge above the surface. The whole system is filled with water and tightly sealed. As roots remove water from the soil, the soil draws water from the porous cup, creating a vacuum that is measured by the gauge. Drier soil or media has a greater suction. Tensiometers are effective for suctions of 0 (saturation) to 0.8 atmospheres, about 80 centibars. Suction values greater than 80 centibars allow air to enter through the fine pores of the ceramic cup, breaking the vacuum, so the tensiometer no longer functions. Tensiometers are well-suited for shallow-rooted, frequently irrigated crops and nursery applications. Their utility is limited in dry and coarse-textured soils.

Electrical resistance blocks consist of one or more pairs of electrodes embedded in a porous material, usually gypsum, ceramic, or fiberglass. The blocks are buried in the root zone with wires leading to the soil surface. The water content of the block is in equilibrium with the surrounding soil moisture content. When the soil is wet, the electrical resistance is low. As the soil and blocks become dry, the

FIGURE 5-4. Using a tensiometer to determine soil moisture.

resistance increases. The resistance is read by specially-designed meters attached to the two lead wires leading from the block. These meters are portable and low in cost. Resistance blocks operate more effectively in the drier range of soil water content; however, they have limited accuracy in very saline or sandy soils. Gypsum blocks must be individually calibrated. Electrical characteristics of gypsum blocks vary from block to block and must be replaced periodically as the gypsum dissolves. Advanced resistance blocks with gypsum embedded in an insoluble, granular matrix and a porous metal sleeve do not have the limitations of gypsum blocks.

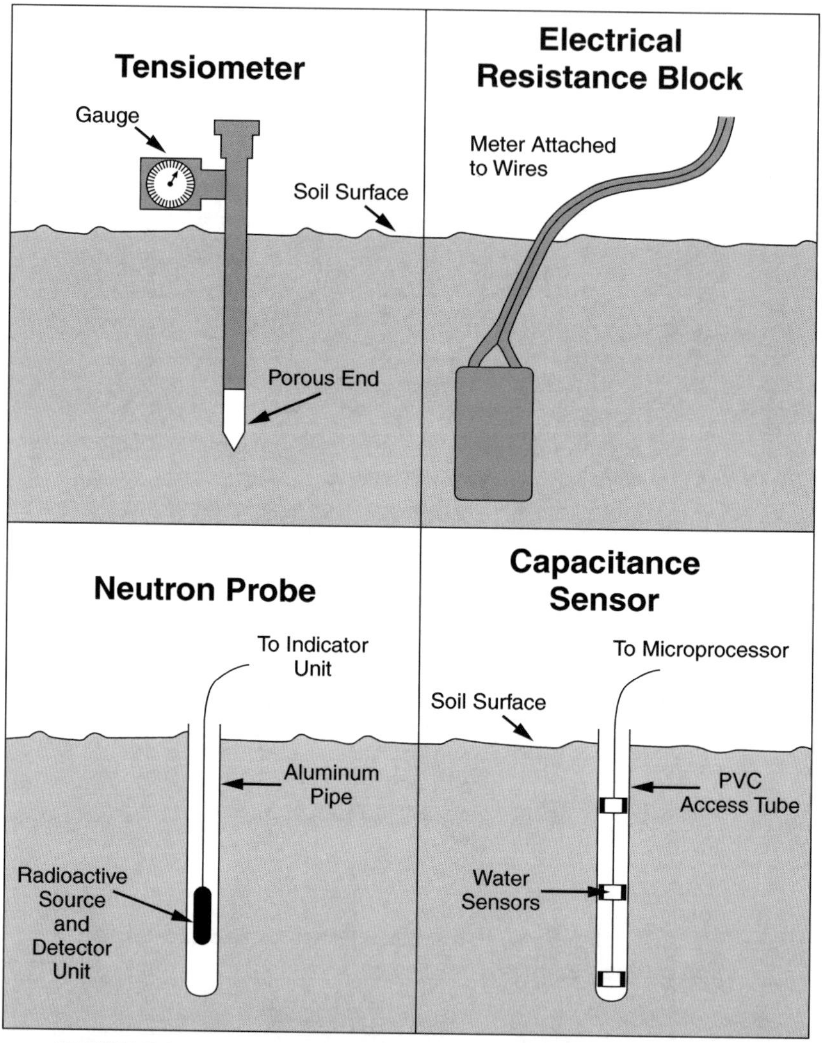

FIGURE 5-5. Examples of four soil moisture measurement devices.

Neutron probes can be used in field situations to determine soil moisture content. A neutron probe contains a radioactive source, a detector tube, and an electronic indicator unit. The source and the detector form a single unit that is lowered into the soil through an access tube. Neutrons emitted by the source slow down as they collide with hydrogen atoms in the soil and return to the detector. The detector counts slow neutrons and the data is displayed or stored. Water is the greatest source of hydrogen in the soil—the wetter the soil, the higher the slow neutron count. Neutron counts are compared with a range of oven-dried soil moisture samples to develop a calibration curve. A calibrated neutron probe gives accurate soil water content data over a wide range of soil moisture contents. The spherical pattern of emitted neutrons reduces the accuracy of soil moisture readings in the top one-half foot of the soil profile. Disadvantages of neutron probes include high cost, need for a trained and licensed operator, disposal fees for the radioactive source, and initial calibration. Neutron probes are not suitable for container-grown plants.

Thermal dissipation sensors indirectly measure soil water content by measuring the dissipation of heat from a porous ceramic block or disk in contact with soil or media. Thermal dissipation sensors are well suited for high-frequency irrigation. The advantage of thermal dissipation sensors is they produce accurate readings of soil moisture independent of soil or media texture, temperature, and salinity. Once the sensor is calibrated, it is useful in any soil. It is also relatively maintenance-free. Their weaknesses include high cost and the need to calibrate each sensor.

There are two types of *capacitance sensors*—sensors that use either time domain reflectometry (TDR) or frequency domain response (FDR) to detect soil moisture by measuring the apparent dielectric constant of the soil-water-air matrix. The dielectric constant is a measure of the ability of any material to transmit an electrical field. The dielectric constant of water is about 80 compared to air that has a dielectric of one and soil minerals that have a dielectric between two and five.

Time domain reflectometer devices (TDR) generate a high-speed, microwave pulse that travels down a set of rigid wires buried in the soil. The velocity of the pulse in the soil is determined primarily by water content. The returning pulse is detected via a microprocessor, and the travel time of the wave is used to calculate the water content of the soil.

Frequency domain response devices (FDR) emit high frequency radio waves into soil surrounding the sensor. Soil moisture changes the frequency of the emitted signal. The ratio of initial to final frequency is correlated to soil water content.

The advantage of both FDR and TDR sensors is the ability to continuously record soil moisture data. Both TDR and FDR devices measure a small volume of soil around the probe. Careful installation is critical to ensure good soil contact with the probe. TDR sensors require no calibration (except where the soil has greater than 10 percent organic matter) and are not sensitive to normal salinity. FDR sensors are more sensitive to changes in soil temperature and salinity than TDR sensors and should be calibrated for greatest accuracy.

The number of instruments needed and the interpretation of the readings depend upon a host of variables including cost of the sensors, soil variability, media components, the placement and installation of sensors, the plants being grown, and the types of irrigation systems used. Measurements of soil moisture do not directly indicate potential plant moisture stress, but are useful for irrigation scheduling.

PLANT MOISTURE MEASUREMENTS

Plants consist mainly of water. Water under pressure keeps plant stems and leaves erect so sunlight can be captured for photosynthesis. Water moves from the soil to the roots, stems, leaves, and stomata, carrying essential nutrients for growth and development. The energy status of water within a plant is called *total plant water potential*. Water moves from high water potential to low water potential. Healthy plants typically have less negative water potentials than plants that are under water stress.

The total plant water potential is a sum of three interacting pressure potential components. *Pressure potential*, or turgor, occurs when water enters a cell with a rigid wall causing the pressure to increase. Pressure potential is usually positive. *Osmotic potential* is caused by the effect sugars, ions, and other dissolved solutes have on water pressure. Osmotic potential is always negative. Finally, *matric potential* is caused by adhesive forces due to the surface tension of water. Plant matric potential is almost always negative.

Pressure bombs are devices used to measure plant water potential. A pressure bomb consists of a pressure chamber, a pressure gauge, a control valve, and compressed nitrogen as a pressure source. The petiole from a freshly cut leaf is extended through a rubber-sealing stopper with the leaf inside the pressure chamber. When the petiole is cut, plant fluids withdraw into the xylem vessels because the pressure outside the plant is greater than inside the conducting tissue. Once the chamber is sealed, pressure is slowly increased until water in the xylem is forced back out to the cut surface. At this point, the positive chamber pressure matches the negative potential of the xylem fluid. A higher gauge reading means a more negative leaf water potential, indicating increasing plant water stress. Disadvantages of the pressure bomb include measurement sensitivity to time of day, time required to take readings, and the lack of leaf water potential guidelines for specific ornamental plants.

Sap flow sensors measure the thermal dissipation of an amount of heat transferred to and carried by plant sap. Sap flow sensors consist of cuffs that are wrapped about the stem of a plant. Advantages of sap flow sensors are the ability to give real-time readings of moisture movement up the stem of the plant that can be recorded or logged electronically. Sap flow sensors also require no calibration to determine sap flow and are available in a wide range of diameters to accommodate many species of plants. The disadvantage is that it can be difficult to extrapolate single stem sap flow measurements to overall plant canopy transpiration. In addition, sap flow sensors may not function well in very hot environments.

Trunk or *stem diameter sensors* use sensitive pressure transducers to detect minute changes in the diameter of a plant stem, which in turn can be related to the plant water content. Trunk diameter sensors are non-invasive and have been shown to give relative indications of changes in plant water status. A disadvantage of this method lies in the difficulty of correlating changes in trunk diameter with the amount of irrigation water necessary to restore the plant to optimum moisture status.

Infrared thermometry techniques indirectly measure plant moisture stress by comparing plant canopy temperatures with ambient temperatures, relative humidity, and sunlight intensity. Since plants warm as they undergo moisture stress due to reductions in transpiration, infrared thermometry measures canopy temperature and presents this information as stress indices. Changes in plant stress indices can help with irrigation scheduling. Infrared thermometers are portable, operate independent of soil conditions, and require little preparation, calibration, or maintenance. Leaf temperatures can be identified over large areas or a single plant. Infrared thermometry is limited by its sensitivity to user technique. Each plant type requires the development of a specific stress index.

Although plant moisture sensing devices are useful tools, they all have inherent limitations that must be thoroughly understood by the user. Further research is needed to refine the techniques and guidelines that will make these tools even more useful.

WATER ANALYSIS TERMINOLOGY

Various terms are used in reporting chemical analyses of water. An understanding of the commonly used reporting methods is needed to interpret the data properly.

Dissolved salts dissociate into electrically charged particles called *ions*. Positively charged ions are called *cations*; negatively charged ions are called *anions*.

Table 5-4 lists the common ions reported in water analyses by laboratories. The common cations are calcium (Ca^{2+}), magnesium

TABLE 5-4

Major Constituents in Irrigation Water

		Symbol	Equivalent Weight
			-----------(g)----------
Cation	Calcium	Ca^{2+}	20
	Magnesium	Mg^{2+}	12
	Sodium	Na^{+}	23
	Potassium	K^{+}	39
Anions	Bicarbonate	HCO_3^-	61
	Carbonate	CO_3^{2-}	30
	Chloride	Cl^-	35.5
	Sulfate	SO_4^{2-}	48

(Mg^{2+}), sodium (Na^{+}), and potassium (K^{+}). Other cations that may be reported are iron (Fe^{2+} or Fe^{3+}) and manganese (Mn^{2+}). The anions usually reported are bicarbonate (HCO_3^-), carbonate (CO_3^{2-}), chloride (Cl^-), and sulfate (SO_4^{2-}). Other anions that may be present in water are borate ($H_2BO_3^-$) and nitrate (NO_3^-), which is sometimes reported as nitrate-nitrogen (NO_3–N).

Three principal units are used to express the concentration of constituents in water: parts per million, milligrams per liter, and milliequivalents per liter.

Milliequivalents per liter (meq/L) is the most useful method of reporting the major chemical components of water and is a measurement of charge concentration per liter. Salts are combinations of cations (sodium, potassium, calcium, magnesium, etc.) and anions (chloride, sulfate, bicarbonate, etc.) in definite weight ratios. These weight ratios are based upon the atomic weight of each constituent and upon the valence (electrical charge). An equivalent weight of an ion is the atomic weight divided by its valence. See Table 5-4 for equivalent weights of the common ions. A simple method of checking the accuracy of a water analysis report is to compare the sum of the cations with the sum of the anions. The sum of cations should roughly equal the sum of the anions. Either sum should roughly equal the electrical conductivity of the water sample (EC_w) × 10.

Parts per million (ppm) is defined as one part of a substance to one million parts of water—or one milligram of a substance per one kilogram of solution. Since one kilogram of water equals one liter, ppm is interchangeable with *milligrams per liter (mg/L)*. The term "parts per million" is used to report constituents found in low concentrations in irrigation water such as iron, manganese, boron, nitrates, and nitrate-nitrogen.

Due to the differences among reports from various laboratories, it is useful to know how to convert from one unit of measurement to another. Parts per million, milligrams per liter, and milligrams per kilogram can be used interchangeably. To convert ppm to meq/L or vice versa, first determine the equivalent weight of the ion in question from the information in Table 5-4, then follow these calculations:

$$\text{ppm} = \text{equivalent weight} \times \text{meq/L}$$
$$\text{meq/L} = \text{ppm} \div \text{equivalent weight}$$

TABLE 5-5

Water Analysis Conversions from meq/L or mg/L (ppm) Pounds per Acre-Foot of Water

Ions	Equivalent Weight (A)	Analysis Result (B)	Conversion Factor (C)	Lb of Material / Acre-Foot of Water (D)[1]	Material
Ca	20.04	1 meq/L	2.72	54.5	Calcium
Mg	12.15	1 meq/L	2.72	33.0	Magnesium
Na	23.00	1 meq/L	2.72	62.6	Sodium
CO_3	30.00	1 meq/L	2.72	81.6	Carbonate
HCO_3	61.02	1 meq/L	2.72	166.0	Bicarbonate
Cl	35.46	1 meq/L	2.72	96.5	Chloride
K		1 ppm	3.26	3.26	Potash (K_2O)
P		1 ppm	6.22	6.22	Phosphate (P_2O_5)
NO_3		1 ppm	0.61	0.61	Nitrate
NO_3–N		1 ppm	2.72	2.72	Nitrate-N
SO_4		1 ppm	0.90	0.90	Sulfur

[1] If ion is reported as meq/L, then A × B × C = D;

Example: 2 meq/L of Ca results in 20.04 × 2 × 2.72 = 109 lbs of calcium per acre-foot of water.

If ion is in ppm, then B × C = D;

Example: 7 ppm of K results in 7 × 3.26 = 23 lbs of potash (K_2O) per acre-foot of water.

Table 5-5 is useful in determining the amounts of dissolved salts applied to land with each acre-foot of irrigation water. This calculation is particularly useful for estimating the plant nutrient inputs derived from irrigation water. For example, if water contains 7 ppm NO_3–N then the amount of N applied per acre-foot = 7 × 2.72 = 19.0 lb N. If 3 acre-feet of water is applied during the season, then a total of 19.0 × 3 = 57 lb N is applied per season from the irrigation water. If nitrate alone is reported (not nitrate-N), multiply the ppm of NO_3 by 0.61 to determine pounds of nitrogen per acre-foot of water.

Acidity or alkalinity of water is expressed as *pH*. A pH reading of less than 7.0 is acidic, 7.0 is neutral, and above 7.0 is alkaline or basic. Most western well waters range from pH 7.0 to pH 8.5. Most Western surface waters have a pH that varies from 6.0 to 8.0 and above.

Total salt content in water is usually reported as *electrical conductivity* (EC_w). Chemically-pure water does not conduct electricity. Most irrigation water contains dissolved salts. Increasing amounts of salt in the water will increase the water conductivity. Electrical conductivity of a water sample is used to estimate its salt content. Electrical conductivity is usually reported as *decisiemens per meter (dS/m)* or *millimhos per centimeter (mmhos/cm)*. Both are numerically equivalent units of measurement.

Total dissolved solids (TDS) is another measure of the total salt content of water and is usually reported as parts per million (ppm). Evaporating all the water from a water sample of known weight and then weighing the remaining salt can be used to determine TDS. However, for speed and convenience TDS is typically estimated by measuring the EC_w in dS/m (or mmhos/cm) and multiplying it by 640. The multiplier actually varies depending on the chemicals in the water, so it is best to regard total salt content in terms of electrical conductivity.

Electrical conductivity is commonly used to check the salt content of soils. The conductivity is measured from a soil saturation paste extract and is designated EC_e. This measurement is used to monitor changes in salt concentration of the soil resulting from irrigation. It is also useful in evaluating the relative tolerance of plants to salt and the suitability of a soil for certain plants. One can estimate the EC_e by multiplying EC_w by 1.5.

SOIL PROPERTIES AND WATER QUALITY

Over time, the quality of irrigation water and irrigation practices affects the characteristics of soil. These potential changes in soil properties may influence plant growth and performance.

The Cations

The major base cations present in most irrigation waters—calcium, magnesium, potassium, and sodium—can have a profound impact on the physical and chemical properties of soils. These cations are called *base cations* since as their presence increases, so does the soil pH. Calcium, magnesium, and potassium are all essential plant nutrients.

Calcium (Ca^{2+}) is found in virtually all natural waters. Most soils and soilless media well-supplied with exchangeable calcium usually permit water to penetrate easily. For these reasons, calcium in the form of gypsum or calcium-based fertilizers may be applied to "tight" soils to improve their physical properties. Calcium replaces sodium on the soil or media particles allowing sodium to be leached below the root zone. Thus, irrigation water containing predominantly calcium is most desirable. However, adding additional calcium to soils already well-supplied with exchangeable calcium will not further improve its physical properties. Adding gypsum to a heavy-textured soil will not markedly improve the porosity unless the soil, which contains calcium, previously contained considerable amounts of sodium.

Magnesium (Mg^{2+}) is also found in measurable amounts in most natural waters. Magnesium behaves most similarly to calcium in the soil, but without the benefits to soil structure. When magnesium concentrations become equivalent or greater than calcium, soil may have poor water infiltration and tilth.

Sodium (Na^{+}) salts are very soluble and are found in most natural waters. A soil with a large amount of sodium has poor properties for plant growth due to the lack of structure and aggregation caused by clay dispersion. When soil dominated by sodium cations is wet, it runs together, becomes sticky, and is nearly impervious to water. When it dries, hard massive clods form making it difficult to till. Continued use of water with a high proportion of sodium may result in these negative effects in an otherwise productive soil. Soilless media permeability is not affected by excess sodium like mineral soils. However, high sodium can lead to toxicity that can burn foliage and interfere with the uptake of other cations. High sodium concentrations are often associated with elevated pH, leading to potential problems with metal micronutrient deficiencies.

Potassium (K^{+}) is usually found in small amounts in natural waters. It is strongly adsorbed by the soil. However, like sodium, it can cause the destruction of soil structure by clay dispersion if it becomes the dominant cation.

THE ANIONS

Anions influence soil properties indirectly by affecting exchangeable calcium and sodium ratios and directly by increasing salinity. The most important anions for water quality are bicarbonate, carbonate, chloride, and sulfate.

Bicarbonate (HCO_3^-) is common in natural waters. Sodium and potassium bicarbonates can exist as solid salts (e.g. baking soda or sodium bicarbonate) or dissolved in water. Calcium and magnesium bicarbonates exist only in solution. As moisture in the soil is reduced, either through transpiration or evaporation, calcium bicarbonate decomposes forming carbon dioxide (CO_2) gas and water (H_2O), which leaves insoluble calcium carbonate behind ($CaCO_3$ or lime).

$$Ca(HCO_3)_2 \xrightarrow{\text{upon drying}} CaCO_3 + CO_2(g) + H_2O$$

Bicarbonate ions in soil solution precipitate calcium as the soil dries. This process removes calcium from the clay, leaving sodium and other cations on the exchange sites. A calcium-dominant soil can become a sodium-dominant (sodic) soil over time if irrigation water high in bicarbonate and sodium is used. Container growers must control excess alkalinity if bicarbonate and carbonates are present in irrigation waters. Alkalinity can raise media pH over time to the point where nutrient deficiencies occur. Excess alkalinity can be controlled by acidifying irrigation water to a pH of 6.5 or below with sulfuric, phosphoric, or other acids.

Carbonate (CO_3^{2-}) is found in some water having a pH greater than 8.0. Since calcium and magnesium carbonates are relatively insoluble, the cations associated with high-carbonate waters are likely to be sodium and, in some cases, potassium. When soil dries, carbonate ions will react with calcium and magnesium from the clay in a similar manner to bicarbonate and a sodic soil will develop. Carbonate and bicarbonate are the major contributors to water alkalinity and largely determine the buffering capacity of water. This will affect how much acid is required to lower the pH. It is difficult to predict how water pH will change when injecting acids. It is always best to titrate irrigation water with a desired acid to

TABLE 5-6
Tolerance Thresholds of Several Ornamentals to Chloride

Plant	Chloride
	--------ppm--------
Begonia	650
Impatiens	700
Pelargonium	880
Petunia	1,000
Primula	450
Saintpaulia	450

determine the correct acid injection rate. Many analytical labs perform water titration as a service.

Chloride (Cl^-) is found in all natural waters. In high concentrations, it becomes toxic to most plants. All common chloride salts are soluble and contribute to the total salt content (salinity) of soils. The chloride concentration must be determined to properly evaluate irrigation water. Chloride can be harmful to all plants when present in concentrations that exceed tolerance levels (Table 5-6). Chlorine is very different than chloride. Even very low concentrations of chlorine can damage plants. Low concentrations of chlorine are often added to municipal water supplies as a disinfectant. However, the level of chlorine in municipal waters is mandated by the EPA to be no more than 0.2 part per million at the tap and is rarely toxic to plants. Chlorine rapidly converts to chloride in the presence of organic matter. The presence of chlorine may be of special concern in hydroponic growing systems.

Sulfate (SO_4^{2-}) is abundant in nature. Sodium, magnesium, and potassium sulfates are readily soluble in water. Calcium sulfate dihydrate (gypsum) has limited solubility (2.4 g/L). Sulfate has no specific characteristic action on soil except it contributes to the total salt content. The presence of soluble calcium will limit the solubility of sulfate since calcium sulfate will precipitate in soil. For example, adding a soluble calcium fertilizer to a soil in the presence of sulfate will result in formation of soil gypsum, which has limited solubility.

Nitrate (NO_3^-) is rarely found in large amounts in natural water. However, even small amounts of nitrate in irrigation water can supply plants with nitrogen. Large amounts of nitrate in water may indicate contamination from natural deposits, animal wastes, sewage, or the misapplication of nitrogen fertilizers and organic residues. The federal drinking water standards require nitrate-nitrogen concentrations to be below 10 ppm. Nitrate has no significant effect on the physical properties of soil.

Boron (B) occurs in water in various anion and molecular forms. The usual range in natural water is from 0.01 to 1.0 ppm. Concentrations greater than this are known, but are usually associated with hot springs or brines. Boron has no measurable effect on the physical properties of soil or on soil salinity in amounts that can be tolerated by plants. Boron is not as readily leached from the soil as chloride or nitrate, but most of it can be removed by repeated leaching irrigations with adequate drainage. A small amount of boron is essential for plant growth, but concentrations above the optimum may be phytotoxic. Some plants are more sensitive to excess boron than others. Plants grown on sandy soils irrigated by water low in boron (less than 0.02 ppm) may develop boron deficiencies.

EVALUATING IRRIGATION WATER

A useful evaluation of irrigation water describes its effect on soils and plant growth. This effect is primarily related to the dissolved salts in the water. Depending upon the amount and kind of salts, different soil problems may develop. See Table 5-7 for guidelines to water quality interpretation. An in-depth discussion of the sodium adsorption ratio (SAR) and its effects on water infiltration can be found in Chapter 6.

A *saline soil* contains soluble salts in such quantities that they interfere with most plant growth. The total salt content of soil, or soil salinity, is measured in terms of electrical conductivity of a saturated paste extract (EC_e). The threshold for a saline soil is an EC_e greater than 4.0 dS/m. One of the hazards of irrigated horticulture is the possible accumulation of soluble salts in the root zone. Some plants tolerate more soil salinity than others, but all plants have an upper limit of tolerance. Most plants are more sensitive to salts during early seedling growth and then

TABLE 5-7

Guidelines for Interpretation of Water Quality for Irrigation

		Degrees of Restriction on Use		
Potential Problem	**Units**	**None**	**Slight to Moderate**	**Severe**
pH			Normal range 6.5-8.4	
Salinity				
EC_w (or)	dS/m	<0.7	0.7-3.0	>3.0
TDS	mg/L	<450	450-2,000	>2,000
Infiltration[1]				
SAR = 0-3 and EC_w		>0.7	0.7-0.2	<0.2
SAR = 3-6 and EC_w		>1.2	1.2-0.3	<0.3
SAR = 6-12 and EC_w		>1.9	1.9-0.5	<0.5
SAR = 12-20 and EC_w		>2.9	2.9-1.3	<1.3
SAR = 20-40 and EC_w		>5.0	5.0-2.9	<2.9
Specific ion effects				
Sodium[2]				
Surface irrigation	SAR	<3	3-9	>9
Sprinkler irrigation	meq/L	<3	>3	
Chloride[2]				
Surface irrigation	meq/L	<4	4-10	>10
Sprinkler irrigation	meq/L	<3	>3	
Boron	mg/L	<0.7	0.7-3.0	>3.0
Bicarbonate[3]	meq/L	<1.5	1.5-8.5	>8.5

[1]At a given sodium adsorption ratio (SAR), infiltration rate increases as EC_w increases. Evaluate the potential infiltration hazard for a given SAR by cross-referencing with the given EC_w. Infiltration restrictions are not an issue in soilless media.

[2]For surface irrigation, most woody plants are sensitive to sodium and chloride; use the values shown. Most annual crops are not sensitive; use salinity tolerance tables. For chloride tolerance of selected crops, see Table 5-6 (page 106). With overhead sprinkler irrigation and low humidity (<30%), sodium and chloride may be absorbed through the leaves of sensitive crops.

[3]Applies to overhead sprinkling only.

become increasingly salt tolerant during later stages of growth and development. Where ample water is used to remove excess salt from the root zone, the measured salt level of the soil saturation extract is about 1.5 times that of the irrigation water. Where water is used more sparingly, there may be three times as much salt in the saturation extract. Refer to Tables 5-8 through 5-16 (pages 110–130) for information on salinity tolerance of turf grass and ornamental plants.

A *sodic (alkali) soil* contains enough sodium adsorbed on the clay particles to interfere with plant growth and water infiltration. The threshold for a sodic soil is an exchangeable sodium percentage (ESP) of greater than 15 percent. If a sodic soil is relatively free of soluble salts, it is called a *nonsaline-sodic soil* or just a *sodic soil*. If, in addition to being sodic, it has sufficient soluble salts to restrict plant growth, it is called a *saline-sodic soil*. Excess sodium rarely presents an infiltration problem in soilless media, but can result in elevated media pH and specific ion toxicity. A more complete explanation of sodic soils and their management can be found in Chapter 6.

SALINITY MANAGEMENT

Soluble salts move with soil water. In areas of flooding or sprinkling, salt movement is generally downward. Upward and horizontal movement and concentration of salts commonly occurs as salts are left behind when water evaporates from the soil or media.

Ordinary irrigation methods result in some leaching so that the accumulation of salts in soil is reduced but not eliminated. Before a critical assessment of the salinity hazard of any irrigation water is made, it is necessary to know how much salt a plant can tolerate and how much leaching is needed to reduce the salt in the soil or growing medium to an acceptable level. The only way to remove salts from the root zone is to leach them out with excess water. There are no amendments that can be added to soil that will remove salt. However, improving the soil or media properties to facilitate water movement will speed the reclamation of saline soils.

Tables 5-8 through 5-16 (pages 110–130) show the estimated tolerance to soil salinity and salt spray for numerous ornamentals and turfgrasses. Though extensive, the nine tables are by no means exhaustive. Refer to the supplementary reading at the end of this chapter for more information.

TABLE 5-8

Salinity Tolerance of Cool and Warm Season Turfgrasses

Cool-season turfgrasses	
Botanical Name	**Common Name**
Tolerant (EC_e > 10 dS/m)	
Puccinellia spp.	alkaligrass
Moderately Tolerant (6 < EC_e < 10 dS/m)	
Agropyron cristatum (L.) Gaertn.	fairway wheatgrass
Agropyron smithii Rydb.	western wheatgrass
Agrostis palustris Seaside	creeping bentgrass cv. Seaside
Festuca arundinacea Schreb.	tall fescue
Festuca rubra L. spp. *trichophylla*	slender creeping red fescue cv. Dawson
Lolium perenne L.	perennial ryegrass
Moderately Sensitive (3 < EC_e < 6 dS/m)	
Agrostis palustris Huds.	creeping bentgrass
Festuca longifolia Thuill.	hard fescue
Festuca rubra L. Spp. *commutata* Gaud.	chewings fescue
Festuca rubra L. spp. *rubra*	creeping red fescue
Lolium multiflorum Lam.	annual ryegrass
Sensitive EC_e < 3 dS/m	
Agrostis tenuis Sibth.	colonial bentgrass
Poa annua L.	annual bluegrass
Poa pratensis L.	Kentucky bluegrass
Poa trivialis L.	rough bluegrass

TABLE 5-8 (Continued)
Warm-season turfgrasses

Botanical Name	Common Name
Tolerant ECe > 10 dS/m	
Cynodon spp.	Bermudagrass
Papsalum vaginatum Swartz.	seashore paspalum
Stenotaphrum secundatum (Walter) Kuntze	St. Augustinegrass
Moderately Tolerant (6 < EC_e < 10 dS/m)	
Bouteloua gracilis (H.B.K.) Lag. ex steud.	blue grama
Buchloë dactyloides (Nutt.) Engelm.	buffalograss
Zoysia spp.	Zoysiagrass
Moderately Sensitive (3 < EC_e < 6 dS/m)	
Paspalum notatum Fluegge	bahiagrass
Sensitive (EC_e < 3 dS/m)	
Eremochloa ophiuroides (Munro) Hackel	centipedegrass

Tanji, K.K. et al. 2007. *Salinity Management Guide.* WateReuse Foundation, Alexandria, VA. www.salinitymanagement.org/Salinity%20Management%Guide/cp/cp_7html.

TABLE 5-9
Salinity Tolerance of Vines, Ground Covers, and Bedding Plants

Botanical Name	Common Name
Highly Tolerant (Ec_e > 6 dS/m)	
Delosperma 'Alba' N. E.	white iceplant
Drosanthemum hispidum Schwantes.	rosea iceplant
Lampranthus productus N. E. Br.	purple iceplant
Malephora crocea Schwantes.	iceplant
Tolerant (4 dS/m < Ec_e < 6 dS/m)	
Allamanda cathartica L.	allamanda

(continued)

TABLE 5-9 (Continued)

Aloe vera Burm. f.	aloe
Aptenia cordifolia N. E. Br.	red apple iceplant
Bougainvillea glabra Choisy	bougainvillea
Carissa macrocarpa A. DC.	natal plum
Carpobrotus edulis L. Bolus.	Hottentot fig
Cuphea hyssopifolia Kunth.	false heather
Ficus pumila L.	creeping fig
Hedera canariensis Willd.	Algerian ivy
Ipomoea pescaprae R. Br.	railroad vine
Ipomoea stolonifera Gmel.	seafoam morning glory
Juniperus conferta Parl.	shore juniper
Juniperus horizontalis Moench.	creeping juniper
Nephrolepis exaltata Schott.	sword fern
Tecomaria capensis Spach.	Cape honeysuckle
Trachelospermum jasminoides Lem.	star jasmine
Tradescantia pallida Hunt.	purple queen
Zamia integrifolia L. f.	coontie
Moderate (2 dS/m < EC_e < 4 dS/m)	
Adiantum sp. L.	maidenhair fern
Allamanda blanchetii A. DC.	purple allamanda
Alternanthera ficoidea R. Br.	joyweed
Antigonon leptopus Hookery	coral vine
Arctostaphylos densiflora 'Lynne' M.S.Back.	Lynne's Vine Hill manzanita
Bromeliaceae sp. L.	bromeliads
Catharanthus roseus G. Donf.	periwinkle
Chlorophytum comosum Jacq.	spider plant
Cyperus altenifolius L.	umbrella sedge

TABLE 5-9 (Continued)

Dietes spp. Salisb. ex Klatt.	African iris
Epipremnum sp. Schott.	pothos
Hedera helix L.	English ivy
Hemerocallis sp. L.	daylily
Hylocereus undatus Britton & Rose	night blooming cereus
Iris hexagona Walter	Iris
Juniperus chinensis L.	Chinese juniper
Juniperus procumbens Siebild ex Endl.	Japanese garden juniper
Kalanchoe sp. Adans.	kalanchoe
Liriope muscari L. H. Bail.	lilyturf (liriope)
Philodendron williamsii Hook.	philodendron
Rosmarinus officinalis L.	rosemary
Tigridia pavonia Ker Gawler	tiger flower
Tulbaghia violacea Harvey	society garlic
Sensitive (EC_e < 2 dS/m)	
Ajuga repens	carpet bugle
Athyrium filix-femina Rith.	lady fern
Caladium sp. Vent.	caladium
Campsis radicans Seem.	trumpet creeper
Clerodendrum thomsoniae Balf. f.	bleeding heart vine
Clytostoma callistegioides Miers ex Bur.	violet trumpet vine
Passiflora incarnata L.	passion flower
Peperomia obtusifolia Dietr.	peperomia
Portulaca grandiflora Hook.	purslane (rose moss)
Salvia farinacea Benth.	mealycup sage
Verbena sp. L.	verbena

Tanji, K.K. et al. 2007. *Salinity Management Guide*. WateReuse Foundation, Alexandria, VA. www.salinitymanagement.org/Salinity%20Management%Guide/cp/cp_7html.

TABLE 5-10
Salinity Tolerance of Shrubs

Botanical Name	Common Name
Highly Tolerant (EC_e > 6 dS/m)	
Euphorbia milii Ch. Des Moulins	crown of thorns
Parthenium argentatum Gray.	guayule
Yucca aloifolia L.	Spanish bayonet
Tolerant (4 dS/m < EC_e < 6 dS/m)	
Acacia redolens Maslin.	prostrate acacia
Agave americana L.	century plant
Arctostaphylos densiflora M.S.Bac	Vine Hill manzanita
Carissa macrocarpa A. DC.	natal plum
Elaeagnus pungens Thunb.	silverthorn, silverberry
Gamolepis chrysanthemoides DC.	African bush daisy
Ilex vomitoria Ait.	yaupon holly
Ilex vomitoria Nana	dwarf yaupon holly
Lantana camara L.	lantana
Myrica cerifera L.	wax myrtle
Myrtus communis L.	true myrtle
Nerium oleander L.	oleander
Opuntia sp. Miller	opuntia cactus
Pittosporum tobira Aiton	mock orange
Plumbago auriculata am.	Cape plumbago
Raphiolepis indica Lindl.	Indian hawthorn
Moderate (2 dS/m < EC_e < 4 dS/m)	
Bambusa sp. Schreb.	bamboo
Buxus microphylla Mull. Arg.	Japanese boxwood
Callistemon rigidus R. Br.	bottle brush

TABLE 5-10 (Continued)

Canna generalis Bailey.	canna lily
Carica papaya L.	papaya
Ceanothus thyrsiflorus Esch.	blue blossom
Cestrum aurantiacum Lindl.	orange cestrum
Dracaena deremensis Engler.	dracaena
Escallonia rubra Pers.	escallonia
Euryops pectinatus L.	golden shrub daisy
Forsythia x intermedia Zabel	hybrid forsythia
Gardenia augusta Merrill	Cape jasmine, gardenia
Heliconia sp.	heliconia
Hibiscus rosa-sinensis L.	rose of China, garden hibiscus
Hydrangea macrophylla Ser.	hydrangea
Ilex cornuta Burford	Chinese Holly
Jasminum polyanthum Franch.	jasmine
Jatropha multifida L.	coral plant
Pyracantha coccinea Roem.	red firethorn
Russelia equisetiformis Schlecht & Cham.	firecracker plant
Sambucus callicarpa Greene	coast red elderberry
Schefflera arboricola L.	dwarf shefflera
Strelitzia reginae Bankses Dryander	bird of paradise
Viburnum odoratissimum Ker.	sweet viburnum
Viburnum suspensum Lindl.	Sandankwa viburnum
Sensitive (EC_e < 2 dS/m)	
Abelia grandiflora Rehd.	abelia 'Edward Goucher'
Acalypha wilkesiana Muell.	copper leaf
Buddleja davidii Franch.	butterfly bush
Calliandra haematocephala Hassk.	powder puff tree
Camellia japonica L.	camellia

(*continued*)

TABLE 5-10 (Continued)

Codiaeum variegatum Blume.	croton
Cornus mas L.	Cornelian cherry
Cotoneaster congestus Baker	Pyrenees cotoneaster
Cotoneaster microphylla Lindl.	Rockspray cotoneaster
Eugenia unifora L.	Surinam cherry
Euphorbia pulcherrima Willd.	poinsettia
Ixora coccinea L.	ixora
Justicia brandegeana Wassh.	shrimp plant
Mahonia aquifolium Nutt.	Oregon grape
Mahonia pinnata Fedde	California holly grape
Murraya paniculata L.	orange jessamine
Nandina domestica Thunb.	heavenly bamboo
Pentas lanceolata Deflers	pentas, Egyptian star-cluster
Photinia fraseri Dress	photinia
Photinia glabra Maxim.	Japanese photinia
Podocarpus macrophyllus D. Don	yew pine
Rosa sp. L.	rose

Tanji, K.K. et al. 2007. *Salinity Management Guide.* WateReuse Foundation, Alexandria, VA. www.salinitymanagement.org/Salinity%20Management%Guide/cp/cp_7html.

TABLE 5-11

Salinity Tolerance of Trees

Botanical Name	Common Name
Highly Tolerant (EC_e > 6 dS/m)	
Parthenium argentatum Gray.	guayule
Tolerant (4 dS/m < EC_e < 6 dS/m)	
Araucaria heterophylla (Salisb.)	Norfolk Island pine

TABLE 5-11 (Continued)

Coccoloba uvifera L.	sea grape
Ficus carica L.	edible fig
Forsythia intermedia Zabel	forsythia
Grevillea robusta Cunn.	silk oak
Juniperus silicicola Bail.	southern red cedar
Juniperus virginiana L.	eastern red cedar, skyrocket juniper
Manilkara *zapota*	sapodilla
Pinus cembroides Zucc.	Mexican piñon pine
Pinus clausa Vasey	sand pine
Plumaria spp. L.	frangipani
Quercus agrifolia Nee	Coast live oak
Quercus virginiana Mill.	live oak
Sapium sebiferum Roxb.	Chinese tallow tree
Moderately Tolerant (2 dS/m < EC_e < 4 dS/m)	
Averrhoa carambola L.	carambola, starfruit
Bauhinia purpurea L.	orchid tree
Callistemon citrinus Curtis.	lemon bottlebrush
Carya illinoiensis Koch.	pecan
Cedrus deodara D. Don	deodar cedar
Cotoneaster microphyllus Lindl.	rockspray or little-leaf cotoneaster
Cupressus sempervirens L.	Italian cypress
Diospyros digyna L.	black sapote
Eriobotrya japonica Lindl.	loquat
Fraxinus oxycarpa Bieb. Ex Willd.	Raywood ash
Koelreuteria paniculata Laxm.	goldenrain tree
Ligustrum japonicum Thunb.	Japanese privet
Persea americana Mill.	avocado

(*continued*)

TABLE 5-11 (Continued)

Pinus elliottii Engelm.	Florida slash pine
Pinus halepensis Mill.	aleppo pine
Pinus thunbergii	Japanese black pine
Platycladus orientalis Franco	Oriental arborvitae
Plumbago auriculata Lam.	Cape plumbago
Prunus spinosa L.	blackthorn
Punica granatum L.	pomegranate
Pyrus spinosa Forssk.	almond-leaved pear
Quercus suber L.	cork oak
Schefflera actinophylla Harms	schefflera, umbrella tree
Sequoia sempervirens Endl.	coast redwood var. Los Altos
Ulmus parvifolia 'Drake'	Chinese elm cv. Drake
Ulmus parvifolia Jacq.	Chinese elm
Sensitive (EC_e < 2 dS/m)	
Acer pseudoplatanus L.	sycamore maple
Acer rubrum L.	red maple
Albizia julibrissin Durazz.	mimosa silk tree
Celtis sinensis Pers.	Chinese hackberry
Citrus limon L.	lemon
Citrus paradisi Macf.	grapefruit
Citrus reticulata Blanco.	tangerine
Citrus sinensis Osbeck.	orange
Cornus mas L.	Cornelian cherry
Diospyros virginiana	persimmon
Euryops pectinatus	golden marguerite
Ginkgo biloba L.	ginkgo
Jacaranda mimosifolia D. Don.	jacaranda

TABLE 5-11 (Continued)

Lagerstoemia indica	crape myrtle
Liquidambar styraciflua	sweetgum
Litchi chinensis Sonn.	lychee
Magnolia grandiflora	southern magnolia
Malus sylvestris Mill.	crab apple
Mangifera indica L.	mango
Musa acuminata Colla.	banana
Olea europaea L.	European olive, olive
Pistacia chinensis Bunge.	Chinese pistache
Prunus armeniaca L.	apricot
Prunus caroliniana Ait.	Carolina laurel cherry
Prunus duclis D.A.Webb.	almond
Prunus persica Batsch	peach
Psidium guajava L.	guava
Quercus laurifolia Michux	laurel Oak
Sequoia sempervirens Endl.	coast redwood var. Aptos Blue
Syzgium jambos Alston	rose apple

Tanji, K.K. et al. 2007. *Salinity Management Guide*. WateReuse Foundation, Alexandria, VA. www.salinitymanagement.org/Salinity%20Management%Guide/cp/cp_7html.

TABLE 5-12

Salinity Tolerance of Palms

Botanical Name	Common Name
Tolerant (EC_e 4 - 6 dS/m)	
Butia capitata Becc.	pindo palm
Chamaerops humilis L.	Mediterranean fan palm
Phoenix dactylifera L.	date palm
Sabal palmetto Lodd.	cabbage palmetto

(*continued*)

TABLE 5-12 (Continued)

Serenoa repens Small	saw palmetto
Washingtonia robusta Wendl.	Mexican fan palm
Moderate (EC_e 2 - 4 dS/m)	
Acoelorrhaphe wrightii Becc.	paurotis palm
Caryota mitis Lour.	fishtail palm
Chrysalidocarpus lutescens Wendl.	areca palm
Nolina recurvata Hemsle	ponytail palm (not a true palm)
Phoenix canariensis Chabaud.	Canary Island date
Phoenix reclinata Jacq.	Senegal date palm
Phoenix roebelinii O'Brien.	pygmy date palm
Rhapis excelsa Henry	lady palm
Syagrus romanzoffiana L.	queen palm

Tanji, K.K. et al. 2007. *Salinity Management Guide.* WateReuse Foundation, Alexandria, VA. www.salinitymanagement.org/Salinity%20Management%Guide/cp/cp_7html.

TABLE 5-13

Salt Spray Tolerance of Vines, Ground Covers, and Bedding Plants

Botanical Name	Common Name
Highly Tolerant (No injury with 600 ppm Na and 900 ppm Cl)	
Aloe vera Burm. f.	aloe
Bougainvillea glabra Choisy	bougainvillea
Carissa macrocarpa A. DC.	natal plum
Carpobrotus edulis L. Bolus.	Hottentot fig
Delosperma 'Alba' N. E.	white iceplant
Drosanthemum hispidum Schwantes.	rosea iceplant
Ficus pumila L.	creeping fig
Hedera canariensis Willd.	Algerian ivy
Ipomoea pescaprae R. Br.	railroad vine

TABLE 5-13 (Continued)

Ipomoea stolonifera Gmel.	seafoam morning glory
Juniperus horizontalis Moench.	creeping juniper
Lampranthus productus N. E. Br.	purple iceplant
Malephora crocea Schwantes.	iceplant
Nephrolepis exaltata Schott.	sword fern
Tecomaria capensis Spach.	Cape honeysuckle
Tradescantia pallida Hunt.	purple queen
Zamia integrifolia L. f.	coontie
Tolerant (No injury with 200 ppm Na and 400 ppm Cl)	
Allamanda cathartica L.	allamanda
Aptenia cordifolia N. E. Br.	red apple iceplant
Catharanthus roseus G. Donf.	periwinkle
Juniperus conferta Parl.	shore juniper
Tigridia pavonia Ker Gawler	tiger flower
Trachelospermum jasminoides Lem.	star jasmine
Moderate (< 10% injury with 200 ppm Na and 400 ppm Cl)	
Adiantum sp. L.	maidenhair fern
Allamanda blanchetii A. DC.	purple allamanda
Alternanthera ficoidea R. Br.	joyweed
Arctostaphylos densiflora 'Lynne' M. S. Back.	Lynne's Vine Hill manzanita
Bromeliaceae sp. L.	bromeliads
Chlorophytum comosum Jacq.	spider plant
Cuphea hyssopifolia Kunth.	false heather
Cyperus altenifolius L.	umbrella sedge
Dietes spp. Salisb. ex Klatt.	African iris
Epipremnum sp. Schott.	pothos

(*continued*)

TABLE 5-13 (Continued)

Hedera helix L.	English ivy
Hemerocallis sp. L.	daylily
Hylocereus undatus Britton & Rose	night blooming cereus
Iris hexagona Walter	iris
Juniperus chinensis L.	Chinese juniper
Juniperus procumbens Siebild ex Endl.	Japanese garden juniper
Kalanchoe sp. Adans.	kalanchoe
Liriope muscari L. H. Bail.	lilyturf (liriope)
Philodendron williamsii Hook.	philodendron
Portulaca grandiflora Hook.	purslane (rose moss)
Rosmarinus officinalis L.	rosemary
Tulbaghia violacea Harvey	society garlic
Sensitive (> 20% injury with 200 ppm Na and 400 ppm Cl)	
Ajuga repens	carpet bugle
Antigonon leptopus Hookery	coral vine
Athyrium filix-femina Rith.	lady fern
Caladium sp. Vent.	caladium
Campsis radicans Seem.	trumpet creeper
Clerodendrum thomsoniae Balf. f.	bleeding heart vine
Clytostoma callistegioides Miers ex Bur.	violet trumpet vine
Passiflora incarnata L.	passion flower
Peperomia obtusifolia Dietr.	peperomia
Salvia farinacea Benth.	mealycup sage
Verbena sp. L.	verbena

TABLE 5-14
Salt Spray Tolerance of Shrubs

Botanical Name	Common Name
Highly Tolerant (No injury with 600 ppm Na and 900 ppm Cl)	
Agave americana L.	century plant
Carissa macrocarpa A. DC.	natal plum
Elaeagnus pungens Thunb.	silverthorn, silverberry
Euphorbia milii Ch. Des Moulins	crown of thorns
Gamolepis chrysanthemoides DC.	African bush daisy
Ilex vomitoria Nana	dwarf yaupon holly
Lantana camara L.	lantana
Myrica cerifera L.	wax myrtle
Nerium oleander L.	oleander
Parthenium argentatum Gray.	guayule
Pittosporum tobira Aiton	mock orange
Raphiolepis indica Lindl.	Indian hawthorn
Yucca aloifolia L.	Spanish bayonet
Tolerant (No injury with 200 ppm Na and 400 ppm Cl)	
Acacia redolens Maslin.	prostrate acacia
Arctostaphylos densiflora M.S.Bac	Vine Hill manzanita
Buxus microphylla Mull. Arg.	Japanese boxwood
Ceanothus thyrsiflorus Esch.	blue blossom
Escallonia rubra Pers.	escallonia
Euryops pectinatus L.	golden shrub daisy
Hydrangea macrophylla Ser.	hydrangea
Ilex vomitoria Ait.	yaupon Holly
Myrtus communis L.	true myrtle

(continued)

TABLE 5-14 (Continued)

Plumbago auriculata am.	Cape plumbago
Sambucus callicarpa Greene	coast red elderberry
Moderate (10% injury with 200 ppm Na and 400 ppm Cl)	
Bambusa sp. Schreb.	bamboo
Callistemon rigidus R. Br.	bottle brush
Canna generalis Bailey.	canna lily
Carica papaya L.	papaya
Cestrum aurantiacum Lindl.	orange cestrum
Cotoneaster microphylla Lindl.	Rockspray cotoneaster
Dracaena deremensis Engler.	dracaena
Forsythia x intermedia Zabel	hybrid forsythia
Gardenia augusta Merrill	Cape jasmine, gardenia
Heliconia sp.	heliconia
Hibiscus rosa-sinensis L.	rose of China, garden hibiscus
Ilex cornuta Burford	Chinese Holly
Jasminum polyanthum Franch.	jasmine
Opuntia sp. Miller	opuntia cactus
Pyracantha coccinea Roem.	red firethorn
Russelia equisetiformis Schlecht & Cham.	firecracker plant
Schefflera arboricola L.	dwarf shefflera
Strelitzia reginae Bankses Dryander	bird of paradise
Viburnum odoratissimum Ker.	sweet viburnum
Viburnum suspensum Lindl.	Sandankwa viburnum
Sensitive (> 20% injury with 200 ppm Na and 400 ppm Cl)	
Abelia grandiflora Rehd.	abelia 'Edward Goucher'
Acacia redolens	spreading acacia
Acalypha wilkesiana Muell.	copper leaf

TABLE 5-14 (Continued)

Atriplex canescens	four-wing saltbush
Baccharis pilularis	coyotebush
Bougainvillea spectabilis	bougainvillea
Buddleja davidii Franch.	butterfly bush
Buxus microphylla	Japanese boxwood
Calliandra haematocephala Hassk.	powder puff tree
Callistemon viminalis	bottle brush
Camellia japonica L.	camellia
Ceanothus thyrsiflorus	wild lilac
Codiaeum variegatum Blume.	croton
Cornus mas L.	Cornelian cherry
Cotoneaster buxifolius	cotoneaster
Cotoneaster congestus	Pyrenees cotoneaster
Cotoneaster congestus Baker	Pyrenees cotoneaster
Elaeagnus pungens	silverberry
Eugenia unifora L.	Surinam cherry
Euonymus japonica	Japanese euonymus
Euphorbia pulcherrima Willd.	poinsettia
Ilex cornuta	Chinese holly
Ilex cornuta 'Burfordii'	Burford holly
Ilex vomitoria	yaupon holly
Ixora coccinea L.	ixora
Jatropha multifida L.	coral plant
Juniperus chinensis	blue point juniper
Juniperus chinensis	Hollywood juniper, spreading juniper
Juniperus sabina 'Buffalo'	buffalo juniper
Justicia brandegeana Wassh.	shrimp plant
Leucophyllum frutescens	Texas sage

(*continued*)

TABLE 5-14 (Continued)

Ligustrum lucidum	glossy privet
Mahonia aquifolium Nutt.	Oregon grape
Mahonia pinnata Fedde	California holly grape
Murraya paniculata L.	orange jessamine
Nandina domestica	nandina
Nandina domestica Thunb.	heavenly bamboo
Olea europaea	European olive
Pentas lanceolata Deflers	pentas, Egyptian star-cluster
Photinia fraseri	red tip photinia
Photinia fraseri Dress	photinia
Photinia glabra Maxim.	Japanese photinia
Pittosporum tobira	dwarf pittosporum
Podocarpus macrophyllus D. Don	yew pine
Pyracantha fortuneana	pyracantha
Raphiolepis indica	Indian hawthorn
Rosa sp. L.	rose
Rosmarinus officinalis	rosemary
Sophora secundiflora	Texas mountain laurel
Thuja orientalis	oriental arborvitae

Tanji, K.K. et al. 2007. *Salinity Management Guide.* WateReuse Foundation, Alexandria, VA. www.salinitymanagement.org/Salinity%20Management%Guide/cp/cp_7html.

TABLE 5-15

Salt Spray Tolerance of Trees

Botanical Name	Common Name
Highly Tolerant (No injury with 600 ppm Na and 900 ppm Cl)	
Araucaria heterophylla (Salisb.)	Norfolk Island pine
Coccoloba uvifera L.	sea grape
Grevillea robusta Cunn.	silk oak

TABLE 5-15 (Continued)

Juniperus silicicola Bail.	southern red cedar
Juniperus virginiana L.	eastern red cedar, skyrocket juniper
Parthenium argentatum Gray.	guayule
Pinus cembroides Zucc.	Mexican piñon pine
Pinus clausa Vasey	sand pine
Quercus virginiana Mill.	live oak
Sapium sebiferum Roxb.	Chinese tallow tree
Tolerant (No injury with 200 ppm Na and 400 ppm Cl)	
Callistemon citrinus Curtis.	lemon bottlebrush
Ficus carica L.	edible fig
Forsythia intermedia Zabel	forsythia
Manilkara zapota	sapodilla
Plumaria spp. L.	frangipani
Plumbago auriculata Lam.	Cape plumbago
Prunus spinosa L.	blackthorn
Quercus agrifolia Nee	Coast live oak
Moderately Tolerant (10% injury with 200 ppm Na and 400 ppm Cl)	
Averrhoa carambola L.	carambola, starfruit
Carya illinoiensis Koch.	pecan
Cedrus deodara D. Don	deodar cedar
Cotoneaster microphyllus Lindl.	rockspray or little-leaf cotoneaster
Cupressus sempervirens L.	Italian cypress
Diospyros digyna L.	black sapote
Eriobotrya japonica Lindl.	loquat
Fraxinus oxycarpa Bieb. Ex Willd.	Raywood ash
Koelreuteria paniculata Laxm.	goldenrain tree

(continued)

TABLE 5-15 (Continued)

Ligustrum japonicum Thunb.	Japanese privet
Persea americana Mill.	avocado
Pinus elliottii Engelm.	Florida slash pine
Pinus halepensis Mill.	aleppo pine
Pinus thunbergii	Japanese black pine
Platycladus orientalis Franco	oriental arborvitae
Prunus caroliniana Ait.	Carolina laurel cherry
Punica granatum L.	pomegrenate
Pyrus spinosa Forssk.	almond-leaved pear
Quercus suber L.	cork oak
Schefflera actinophylla Harms	schefflera, umbrella tree
Sequoia sempervirens Endl.	coast redwood var. Los Altos
Ulmus parvifolia 'Drake'	Chinese elm cv. Drake
Ulmus parvifolia Jacq.	Chinese elm
Sensitive (> 20% injury with 200 ppm Na and 400 ppm Cl)	
Acer pseudoplatanus L.	sycamore maple
Acer rubrum L.	red maple
Albizia julibrissin Durazz.	mimosa silk tree
Bauhinia purpurea L.	orchid tree
Celtis sinensis Pers.	Chinese hackberry
Citrus limon L.	lemon
Citrus paradisi Macf.	grapefruit
Citrus reticulata Blanco.	tangerine
Citrus sinensis Osbeck.	orange
Cornus mas L.	Cornelian cherry
Diospyros virginiana	persimmon
Euryops pectinatus	golden marguerite
Ginkgo biloba L.	ginkgo

TABLE 5-15 (Continued)

Jacaranda mimosifolia D. Don.	jacaranda
Lagerstoemia indica	crape myrtle
Liquidambar styraciflua	sweetgum
Litchi chinensis Sonn.	lychee
Magnolia grandiflora	southern magnolia
Malus sylvestris Mill.	crab apple
Mangifera indica L.	mango
Musa acuminata Colla.	banana
Olea europaea L.	European olive, olive
Pistacia chinensis Bunge.	Chinese pistache
Prunus armeniaca L.	apricot
Prunus duclis D.A.Webb.	almond
Prunus persica Batsch	peach
Psidium guajava L.	guava
Quercus laurifolia Michux	laurel Oak
Sequoia sempervirens Endl.	coast redwood var. Aptos Blue
Syzgium jambos Alston	rose apple

Tanji, K.K. et al. 2007. *Salinity Management Guide.* WateReuse Foundation, Alexandria, VA. www.salinitymanagement.org/Salinity%20Management%Guide/cp/cp_7html.

TABLE 5-16

Salt Spray[1] Tolerance of Palms

Botanical Name	Common Name
Tolerant (No injury with 200 ppm Na & 400 ppm Cl)	
Butia capitata Becc.	pindo palm
Chamaerops humilis L.	Mediterranean fan palm
Phoenix dactylifera L.	date palm
Sabal palmetto Lodd.	cabbage palmetto

(continued)

TABLE 5-16 (Continued)

Serenoa repens Small	saw palmetto
Washingtonia robusta Wendl.	Mexican fan palm
Moderate (10% injury with 200 ppm Na & 400 ppm Cl)	
Acoelorrhaphe wrightii Becc.	paurotis palm
Caryota mitis Lour.	fishtail palm
Chrysalidocarpus lutescens Wendl.	areca palm
Nolina recurvata Hemsle	ponytail palm (not a true palm)
Phoenix canariensis Chabaud.	Canary Island date
Phoenix reclinata Jacq.	Senegal date palm
Phoenix roebelinii O'Brien.	pygmy date palm
Rhapis excelsa Henry	lady palm
Syagrus romanzoffiana L.	queen palm

[1]Salt spray refers to fine aerial droplets of saline sea or irrigation water.

Wu, L., and L. Dodge. 2005. Lanscape Plant Salt Tolerance Selection Guide for Recycled Water Irrigation. A Special Report for the Elvenia J. Slosson Endowment Fund. Dept. of Plant Sci. UC Davis, Davis, CA.

Reasonable irrigation practices should not cause salinity problems if the irrigation water has an EC_w of less than 0.75 dS/m. More salinity problems can be expected with water conductivities between EC_w 0.75 and 3.0 dS/m. Irrigation water with an EC_w greater than 3.0 is likely to impede the growth and development of most plants except for a few salt-tolerant varieties.

The *leaching requirement* is that fraction of water in excess of plant water use necessary to leach salts below the root zone, thereby maintaining average root zone salinity below the plant tolerance level. The leaching requirement must be calculated when irrigating soils with saline water. To determine the leaching requirement, you must know the salinity of the applied irrigation water (EC_w) and the salinity tolerance of the plant (EC_e) as determined from Tables 5-8 to 5-12 (pages 110–120). Note that even when irrigating with water that has salinity below the threshold of a plant, a leaching requirement is necessary to prevent salt buildup. The leaching requirement is calculated as such:

$$LR = \frac{EC_w}{5(EC_e) - EC_w}$$

The following is an example of a leaching requirement calculation: a rose with a salinity tolerance of 3.5dS/m is irrigated with water whose salinity is 2.9 dS/m. Its leaching requirement is $(LR = 2.9 / (5 \times 3.5 - 2.9) = 0.2$.

The following equation is used to calculate the total irrigation amount when including a leaching requirement:

$$AW = \frac{ET}{1 - LR}$$

Where AW is the total depth of water (in inches) to apply, ET is the depth of water needed to meet evapotranspiration, and LR is the leaching requirement as calculated above. *Evapotranspiration* is a measure of the combined depth of water transferred to the atmosphere from the soil or media surface (evaporation) and from the plant (transpiration). Evapotranspiration data can be obtained from county extension services, university publications, and other references on irrigation management. For the rose example above, if the evapotranspiration (ET) is 0.2 inches and the leaching requirement (LR) is 0.2, then the total applied water is 0.25 inches.

Salinity is usually not a problem in modern container nurseries. But where irrigation practices and water quality are poor, salt accumulation may be a problem. Under these conditions, accumulation would occur as pictured in Figure 5-6b. Salt accumulation may also occur where single low-volume emitters do not deliver enough water to wet the entire soil or growing-medium (Figure 5-6c). Nurseries and greenhouses that capture and recycle excess irrigation water to prevent off-site water runoff can develop saline conditions without careful monitoring and management.

Containerized nursery and greenhouse production poses many challenges when dealing with salinity management. Current water quality regulations demand that container nursery and greenhouse operations prevent salt and nutrient-laden irrigation water run-off from exiting the production property. However, most

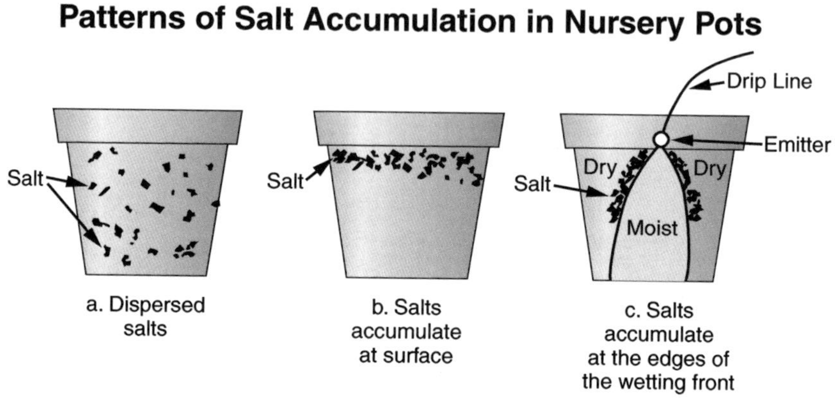

FIGURE 5-6. Water management changes the pattern of salt distribution in containers with a) proper leaching, b) no leaching, and c) drip irrigation.

ornamental plants benefit from an over-application of irrigation water to leach excess salts from soilless media. The current suggested leaching requirement that is recommended for containerized nursery and greenhouse production is 15 percent (LR = 0.15) or less. Irrigation water should be free of all fertilizer when leaching. It is beyond the scope of this handbook to recommend a comprehensive strategy for nursery and greenhouse operations. Consult with qualified irrigation management experts for strategies to limit runoff while maintaining quality plant production.

There are some situations where very pure snowmelt waters are used for irrigation. These waters are very low in total salts, especially calcium. Snowmelt water will leach calcium and other cations from soils. Low soil calcium levels may result in dispersed clay and silt particles leading to water penetration problems. Addition of calcium-containing fertilizers, amendments, wetting agents, or surfactants can help overcome infiltration problems when irrigating with snowmelt water. Snowmelt water does not affect drainage in containers containing soilless media. However, low salt water, including distilled and de-ionized water, lacks a buffering system that can counteract acidification of media by ammoniacal fertilizers. Addition of limestone in the media pre-mix, as well as adding potassium carbonate to the irrigation water, prevents a rapid pH drop when irrigating media with low salt water.

Various devices are sold that claim to reduce salinity. There is often no independent research data for many of these devices. One should proceed cautiously when considering investing in a device that does not have the support of credible, independent data.

TOXIC CONSTITUENTS

Several inorganic constituents, such as boron, chloride, and sodium may be found in natural water at levels that can be toxic to plants. Additionally, high levels of bicarbonate in water have been shown to induce iron deficiencies in some plants. This problem is minor when compared to the role of bicarbonates in creating permeability problems.

Boron Hazard

A small amount of boron is necessary for plant growth. To provide an adequate supply of this plant nutrient, 0.02 ppm boron or more in the irrigation water may be required. Some water contains an adequate supply of boron, but water from many rivers may be deficient. Some well water and a few surface streams contain an excess of boron, thus creating a hazard. Plants grown in soils high in lime may tolerate more boron than those grown in non-calcareous soils. Table 5-17 presents the relative tolerance of plants to boron.

TABLE 5 – 17

Boron Tolerance Limits for Ornamentals

VERY SENSITIVE		
Common Name	**Botanical Name**	**Threshold mg/L**
Oregon grape	Mahonia aquifolium	<0.5
Photinia	Photinia x fraseri	<0.5
Xylosma	Xylosma congestum	<0.5
Thorny elaeagnus	Elaeagnus pungens	<0.5
Laurustinus	Viburnum tinus	<0.5
Wax-leaf privet	Ligustrum japonicum	<0.5
Pineapple guava	Feijoa sellowiana	<0.5

(continued)

TABLE 5-17 (Continued)

Spindle tree	Euonymus japonica	<0.5
Japanese pittosporum	Pittosporum tobira	<0.5
Chinese holly	Ilex cornuta	<0.5
Juniper	Juniperus chinensis	<0.5
Yellow sage	Lantana camara	<0.5
American elm	Ulmus Americana	<0.5

SENSITIVE

Common Name	Botanical Name	Threshold mg/L
Zinnia	Zinnia eleganus	0.5-1.0
Pansy	Viola tricolor	0.5-1.0
Violet	V. odorata	0.5-1.0
Larkspur	Delphinum sp.	0.5-1.0
Glossy abelia	Abelia × grandiflora	0.5-1.0
Rosemary	Rosmarinus officinalis	0.5-1.0
Oriental arborvitae	Platycladus orientalis	0.5-1.0
Geranium	Pelargoium × hortorum	0.5-1.0

MODERATELY SENSITIVE

Common Name	Botanical Name	Threshold mg/L
Gladiolus	Gladiolus sp.	1.0-2.0
Marigold	Calendula officinalis	1.0-2.0
Poinsettia	Euphorbia pulcherrima	1.0-2.0
China aster	Callistephus chinensis	1.0-2.0
Gardenia	Gardenia sp.	1.0-2.0
Southern yew	Podocarpus marcophyllus	1.0-2.0

TABLE 5-17 (Continued)

Brush cherry	Syzygium paniculatum	1.0-2.0
Blue dracaena	Cordyline indivisa	1.0-2.0
Ceniza	Leucophyllus frutescens	1.0-2.0

MODERATELY TOLERANT

Common Name	Botanical Name	Threshold mg/L
Bottlebrush	Callistemon citrinus	2.0-4.0
California poppy	Eschscholzia californica	2.0-4.0
Japanese boxwood	Buxus microphylla	2.0-4.0
Oleander	Nerium oleander	2.0-4.0
Chinese hibiscus	Hibiscus rosa-senensis	2.0-4.0
Sweet pea	Lathyrus odoratus	2.0-4.0
Carnation	Dianthus caryophyllus	2.0-4.0

TOLERANT

Common Name	Botanical Name	Threshold mg/L
Indian hawthorn	Raphiolepis indica	6.0-8.0
Natal plum	Carissa grandiflora	6.0-8.0
Oxalis	Oxalis bowiei	6.0-8.0

Maas. 1990. *Boron Tolerance Limits for Ornamentals.* U.S. Salinity Laboratory Agricultural Research Service. www.ussl.ars.usda.gov/pls/caliche/BORT47.

Chloride Hazard

Chloride ions are found in virtually all natural waters. The chloride anion is soluble and moves freely through soil. Relatively small amounts of chloride are necessary for plant growth. However, high concentrations of chloride are toxic to plants, particularly woody species. Most annuals and short-lived perennials are moderately tolerant of chloride, whereas trees, vines, and woody ornamentals tend to be more sensitive (see Table 5-6, page 106). When present in excessive concentrations, chloride is often the first anion to produce recognizable symptoms such as yellow leaf margins and tips.

Sodium Toxicity

Trees, vines, and woody ornamentals tend to be especially sensitive to excessive sodium. Annual plants are usually not affected directly by high sodium. However, water that has sodium concentrations of greater than 3 meq/L is likely to cause foliar injury symptoms when applied by sprinkler irrigation. The ratio of sodium compared to calcium and magnesium (instead of the absolute concentration) is most critical for plant growth. More information on the effects of high sodium on soil properties can be found in Chapter 6.

WATER FOR SPRINKLER IRRIGATION

Most water suitable for surface irrigation may be safely used for overhead sprinkler irrigation. There are, however, some exceptions. Leaf burn caused by sodium and chloride absorption may occur during periods of high evaporation. Low humidity, high temperature, and wind can increase the concentration of these ions on the leaves as the water dries. Sometimes this can be corrected by increasing the frequency of irrigation, but if this is impractical, it may be necessary to irrigate only at night. Usually there is no problem when irrigation water contains 3 meq/L or less of either sodium or chloride.

Certain plants are sensitive to foliar-applied salts because they absorb sodium and chloride very readily through their foliage. Other plants that are very sensitive to soil salinity absorb salts very slowly through their foliage and are tolerant of salts applied to the foliage. Water used for misting systems should have low salinity to prevent injury from salt spray.

Bicarbonate ions in water can also be a problem with overhead sprinkler irrigation. A white deposit of calcium carbonate may form on the leaves and fruit. This layer of calcium carbonate can render ornamentals unmarketable because they are unattractive; however, this coat of "whitewash" is not known to have an adverse effect on plant growth. Bicarbonate levels below 1.5 meq/L should not cause a problem. Water with bicarbonate concentrations greater than 1.5 meq/L can be acidified to a pH of between 5.0 and 6.5 resulting in the elimination of more than half of the bicarbonate ions.

LOW-VOLUME IRRIGATION

Low-volume irrigation is the frequent, slow application of water through various types of emitters. The most common forms of low-volume irrigation are drip and microsprinkler. System design, components, and flow rate requirements vary according to plant, soil, water, and environmental conditions. For example, a mature tree may require from two to six emitters, with each emitter applying ½ to 2 gallons per hour. Small to medium containers may require only one drip emitter or micro-sprinkler per container (Figure 5-7).

Low rates of water application involve small orifices and need filtered water free of particles. Dissolved salts in the water may crystallize around and in the orifices, and reduce their flow rates or plug them completely. The most common cause of chemical plugging in low-volume irrigation systems is the formation of insoluble calcium carbonate, also known as scale. Other causes of plugging include chemical or microbial oxidation of iron or manganese, bacterial or algal growth, suspended solids, and the reaction of injected fertilizers with ions present in the irrigation water. One should never leave chemicals in low-volume irrigation lines after irrigation has occurred. Fertilizers left in lines will stimulate the growth of bacteria and algae that can plug emitters. Regular flushing of irrigation lines will remove sediment and contaminants

FIGURE 5-7. Nursery container plant with an irrigation stake.

TABLE 5-18
Plugging Potential of Irrigation Water Used in Drip Irrigation Systems

	Potential Restrictions on Use		
	Little	**Moderate**	**Severe**
Suspended solids *(ppm)*	<50	50–100	>100
pH	< 7	7–8	>8
Dissolved solids *(ppm)*	<500	500–2,000	>2,000
Manganese *(ppm)*	<0.1	0.1–1.5	>1.5
Iron *(ppm)*	<0.1	0.1–1.5	>1.5
Hydrogen sulfide *(ppm)*	<0.5	0.5–2.0	>2.0
Bacterial populations (no./mL)	$<10^4$	10^4–5×10^4	$>5 \times 10^4$

that accumulate and could plug emitters. Careful attention to water quality tests is required in order to evaluate the plugging potential of irrigation water. Table 5-18 presents the plugging potential of water used for low-volume irrigation.

Low-volume irrigation is widely used on container-grown plants. Since it may save water, it is also well adapted to areas where water is costly or scarce. The total amount of water used in low-volume irrigation is usually less than in conventional irrigation systems because:

- Runoff is reduced
- Evaporation is reduced
- The total volume of soil or media wetted is usually less
- Deep percolation of water may be reduced
- The time period of water application can be carefully controlled

DRAINAGE

Improving soil drainage may be necessary in conditions with high water tables or where salinity is a hazard. Most crops grow best where the water table is more than 6 feet below the soil surface.

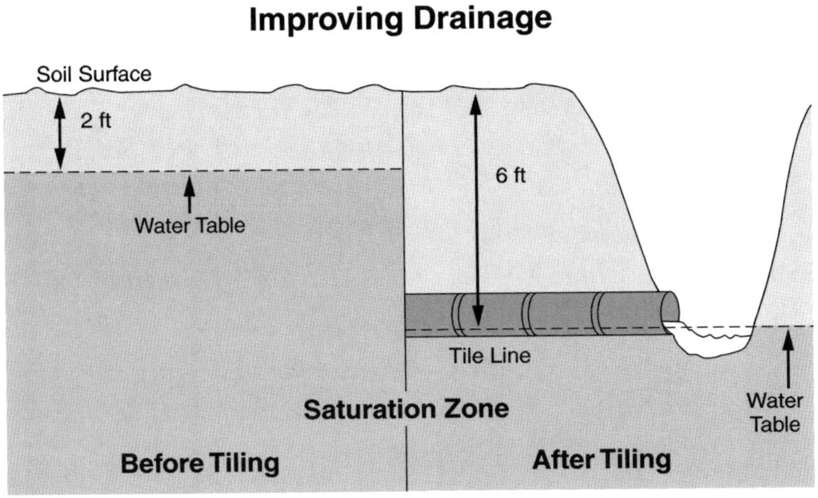

FIGURE 5-8. Tile drainage is used to lower the water table.

A drainage system (Figure 5-8) should be installed in fields where water stands within 6 feet of the surface. Drains remove water if properly placed below the water table or in the saturation zone.

Proper drainage is also essential for plants grown in containers in a soilless media. It is important to apply more water than the substrate can hold; generally 10 to 15 percent of the water applied to the pots should run out the bottom. Without adequate drainage, water will accumulate in the pot and damage the root system.

SUPPLEMENTARY READING

Ayers, R.S., and D.W. Westcott. 1985. *Water Quality for Agriculture*. FAO Irrigation and Drainage Paper No. 29. Rev. 1. Rome, Italy.

Handreck, K., and N. Black. 2007. *Growing Media for Ornamental Plants and Turf.* University of New South Wales Press Ltd., Sydney, Australia.

James, D.W., R.J. Hanks, and J.J. Jurinak. 1982. *Modern Irrigated Soils*. Wiley-Interscience Publication, New York, NY.

Kramer, P.J. 1983. *Water Relations of Plants*. Academic Press, Inc., New York, NY.

Mass, E.V. 1984. "Salt Tolerance of Plants." In *Handbook of Plant Science.* B.R. Christie, ed. CRC Press, Inc., Boca Raton, FL.

Oster, J.D., M.J. Singer, A. Fulton, W. Richardson, and T. Prichard. 1992. *Water Penetration Problems in California Soils: Prevention, Diagnosis and Solution.* Kearney Foundation of Soil Science, Davis, CA.

Reed, D.W. 1996. *Water, Media, and Nutrition for Greenhouse Crops.* Ball Publishing, Batavia, IL.

Rhoades, J.D. 1972. Quality of water for irrigation. *Soil Sci.* 113: 277–284.

Schwankl, L., B. Hanson, and T. Prichard. 1996. *Microirrigation of trees and vines.* University of California, Publication 3378.

Soil Survey Staff. 1954. *Diagnosis and Improvement of Saline and Alkali Soils.* USDA Agricultural Handbook 60. U.S. Gov. Print. Office Washington, DC.

Tanji, K.K. 1990. *Agricultural Salinity Assessment and Management.* ASCE Manuals and Reports on Engineering, Practice No. 71. ASCE Publications, New York, NY.

Tanji, K.K. et al. 2007. *Salinity Management Guide.* WateReuse Foundation, Alexandria, VA. www.salinitymanagement.org/Salinity%20Management%Guide/cp/cp_7.html.

U.S. Salinity Laboratory Staff. 1954. *Diagnosis and Improvement of Saline and Alkali Soils.* USDA Agriculture Handbook No. 60. U.S. Government Printing Office, Washington DC, WA. www.ars.usda.gov/Services/docs.htm?docid=10158&page=2.

Weinhold, F., and H.C. Scharpf. 1997. Tolerance of ornamental plants to salt, sodium and chloride in potting substrates containing compost made of separately collected organic residues. p. 221-228. *On Growing Media and Plant Nutrition.* Proc. Int'l Symp.

Chapter 6

Soil Amendments

Soil is very similar to the foundation of a building. The question to be asked is, "Will this soil support good plant growth?" If not, the soil will need to be amended. A soil *amendment* is something that is added to the soil to make it more productive. This chapter will focus on amending soil to improve plant growth. Amending may be done prior to or after planting. When a soil is adequately supplied with air, water, organic matter, and nutrients, little amending is necessary. Such an ideal, naturally occurring soil is rare. It is often necessary to add an amendment to improve the physical or chemical condition of the soil. Choosing which materials to use is based on the cost, consistency, availability, pH, moisture, and the nutrient-holding capacity of the components. Replacing existing soil with good topsoil is generally more costly than amending. Also, unless careful measures are followed, soil layering often results, which can restrict air and water movement and impede proper root growth and development.

It is essential that the physical and chemical properties of the soil are considered when choosing amendments for optimum results. Forecasting the needs of plants before planting allows the grower to take a proactive approach to plant health. Selecting and incorporating proper amendments prior to planting is very important. It is difficult to amend soils after turf, trees, and shrubs have been planted. This chapter will discuss two types of soil amendments — physical and chemical.

PHYSICAL SOIL AMENDMENTS

The physical properties of soil are related to its texture, as illustrated in Table 6-1. Sandy soils have properties that are different from clay soils. For example, permeability or porosity is high in sandy soils and low in clay soils. Although these differences are recognized in maintenance practices, often the differences are overlooked at the time of amending or planting. Forecasting the needs of plants before planting allows the grower to take a proactive approach to plant health. A *physical amendment* is defined as any substance used for the purpose of promoting plant growth or improving the quality of plants by conditioning soils solely through physical means. Conditioning soils centers primarily on improvements related to water retention, water permeability, and aeration.

Organic Amendments

Organic amendments are derived from living sources, including animals or plants. Organic amendments improve soil properties by physically separating soil particles and by increasing the nutrient and water-holding capacities of the soil. The stable decomposition residue from organic amendments is called *humus*.

Composts

Composts are made by the decomposition of plant residues (Figure 6-1). They usually have good nutrient-holding capacities and high biological activity. Well decomposed composts aid in creat-

TABLE 6-1

Soil Physical Properties Related to Soil Texture

Soil Texture	Permeability	Water Retention
Sand	High	Low
Loam	Medium	Medium
Silt	Low	High
Clay	Low	High

FIGURE 6-1. A commercial composting operation.

ing a desirable soil structure. Mushroom compost is a by-product of mushroom growing beds consisting of soil, straw, manure, wood or peat, and agricultural minerals. Yard waste composts are produced by municipal green waste programs and individual homeowners, but the lack of consistency is a limitation. Composts contribute nutrients and humus to soil mixes.

Composted Manures

Steer, dairy, and poultry manures are best used when fully composted. Composting reduces odor problems and weed seed contamination. Composted manures have fair water and nutrient-holding capacities. Composted manures are sometimes used as a seed cover in new turf seeding or when over-seeding existing turf. Manures have limited use because of their inconsistent nutrient availability and their tendency to contain excess salts.

Wood Residues

Various grades or sizes of wood residues are widely used in amending turfgrass and landscape soils and in preparing container soil mixes (Figure 6-2). These amendments may provide good water

FIGURE 6-2. Bulk redwood soil amendment in storage.

infiltration and oxygen diffusion into soils. However, larger wood residues may possess only poor to fair water and nutrient-holding capacities. In contrast, sawdust and fine wood residues may exhibit excessive water-holding capacities. The rate of decomposition varies with the size and kind of wood residue used. Residues from soft pines may have a residual period of only several months and should be used only when fully composted since they immobilize nitrogen during decomposition. Residues from fir, redwood, cedar, and cypress may last up to five years, decomposing slower than pines.

When wood residues are incorporated into a soil mixture, additional nitrogen must also be incorporated to prevent nitrogen draft (immobilization). This will preclude nitrogen starvation of plants growing in amended soils. Ground fir bark has proven to be a valuable amendment in the turfgrass, landscape, and container industries. Fir bark has low nutrient-holding and fair water-holding capacities. Additionally, the round shape of the bark provides good water and oxygen diffusion rates into soil and resists compaction.

Digested Sludge

Sludge from sewage processing is produced after undergoing about fourteen days of aerobic and anaerobic digestion. After digestion, the sludge is pumped into basins and allowed to dry. Unless further modified, digested sludge is not generally suitable as a soil amendment due to possible contamination with heavy metals, salts, or pathogens. The value of sludge is realized when it is used as an organic base for fertilizers or further composted. Regulatory agencies should be consulted before digested sludge is used on areas such as athletic fields, parks, golf courses, or soil intended for food production.

Composted Sludge

Composted sludge is made by removing water from sludge. Sometimes wood residue or ash is added to sludge. The sludge is then windrowed with biologically active, dried sludge and turned daily for about 40 days. The resulting well-decomposed product has characteristics similar to humus. When used in silt and clay soils, composted sludge improves soil structure. When used in sandy soils, composted sludge improves moisture-holding capacity and biological activity.

Physical Amendment Integrity

Organic compounds from plant residues are not indestructible, although some residues are very resistant to decomposition. The longevity of amendments in the soil depends on soil microbial activity. Microbial decomposition rates depend on soil conditions such as aeration, moisture, temperature, and available nutrients. Most cellulose residues will decompose within 6 months, but lignin residues may last in soil for years. Examples of some organic residues and their relative decomposition rates are found in Table 6-2.

TABLE 6-2

Relative Decomposition Rates of Organic Residues

Days to Weeks	Months	Years
Grass clippings	Leaf molds	Rice hulls
Manures	Composts	Tree bark
Mushroom composts	Humus-type composts	Coir

Selecting Physical Amendments

Table 6-3 presents the soil amending properties of some commonly used amendments. A single material may not have all the desired amending characteristics, so combinations of amendments are often used. Sandy soils benefit from the use of organic materials to increase water retention and nutrient-holding capacity. Fine textured clay and silt soils, by contrast, need fibrous materials to increase permeability and aeration. Organic amendments may be needed to improve water and nutrient retention if the soil mix is low in organic matter content.

Determining Required Quantities of Physical Amendments

When choosing amendments, consider whether they will be effective, practical, and economical. It is also important to consider the plants to be grown, and the size and value of the area to be amended. Generally, it is ineffective to amend a soil using less than 25 percent by volume of an amendment. Heavy use, high value areas, and fine-textured soils with poor structure may benefit from

TABLE 6-3

Soil-Amending Properties of Some Commonly Used Materials

	Amendment	Permeability	Water Retention
Fibrous	Peat	Low to Medium	Very High
	Wood residues	High	Low to Medium
	Ground fir bark	High	Low to Medium
	Rice hulls	High	Low to Medium
Non-Fibrous or Humus	Composts	Low to Medium	Medium to High
	Composted manures	Low to Medium	Medium
	Composted sludge	Low	High
Inorganic	Calcined clay	High	High
	Pumice	High	Low to Medium
	Vermiculite	High	High
	Perlite	High	Low
	Sand	High	Low

TABLE 6-4
Organic Amendment Needed Based on Soil Texture

Texture	% Amendment
Sand	35
Loamy sand	30
Sandy loam	30
Sandy clay loam	25
Sandy clay	25
Loam	25
Silt loam	30
Silty clay loam	30
Clay loam	30
Silt	35
Silty clay	35
Clay	35

the addition of up to 40 to 50 percent of an amendment by volume. Some general guidelines in determining the quantity of organic amendments needed for various soil textures are given in Table 6-4.

Turfgrass

For turfgrass, the general rule is to amend the soil before planting with about 33 percent by volume of an organic amendment. For heavily used areas and for fine-textured soils, the pre-plant amended amount may be increased to 40 percent by volume. For loam soils or on soils where little traffic is expected, this amount may be reduced to a minimum of 25 percent by volume. Table 6-5 may be used to determine the volume of soil amendment needed to amend the soil to various depths.

Bedding Plants, Ground Covers, and Vegetable Gardens

Planting beds, groundcover areas, and vegetable gardens are usually high-volume, low-acreage areas that may be amended by adding 2 inches of amendment over a previously cultivated soil. If the grower incorporates an amendment to a 6 inch depth, the soil should be amended at 33 percent by volume.

TABLE 6-5
Volume of Organic Soil Amendments to Add for Various Depths of Treatment

% Amendment	Depth of Amended Soil in Inches (cubic yards per 1,000 square feet)						
	3 in	*4 in*	*5 in*	*6 in*	*7 in*	*8 in*	*9 in*
5	0.46	0.61	0.77	0.93	1.08	1.23	1.39
10	0.93	1.23	1.54	1.85	2.16	2.47	2.78
20	1.85	2.47	3.09	3.71	4.32	4.94	5.55
30	2.78	3.70	4.64	5.56	6.48	7.41	8.33
40	3.70	4.94	6.18	7.41	8.64	9.88	11.13
50	4.63	6.17	7.72	9.26	10.80	12.34	13.88

Backfill Around Landscape Shrubs and Trees

When planting landscape shrubs and trees, soil amendments are commonly mixed with the soil removed from the planting hole to create *backfill.* The planting hole size should be twice the width and twice the depth of the container the plant is in. The recommended amendment percentage is between 33 and 50 percent. Table 6-6 indicates the quantity of amendments to be added to provide either a 33 or 50 percent mixture by volume for various containers sizes.

Incorporation of Soil Amendments

A uniform soil should be provided throughout the entire root zone. Improper mixing of materials will produce layers or pockets, resulting in interfaces that can restrict air, water, fertilizer, and root movement through the soil. Soil physical amendments are mixed with the existing soil on-site. The exception would be potting soils or specialty growing media that are mixed off-site or on pads or benches. On-site mixing involves uniformly distributing the desired amendment in thin layers over the surface of the soil, followed by thorough mixing with cultivating equipment. This procedure ensures that no layers or pockets remain. Rototillers, turning cultivators, discs, and harrows are often used for mixing. Note that discs tend to leave pockets of unincorporated amendments after

TABLE 6-6

Volume of Organic Soil Amendments Required to Amend Backfill Soils by 33 to 50 Percent for Various Container Sizes

Container Size	1 gal		5 gal		7 gal		15 gal		20 in		24 in	
Amendment (%)	33	50	33	50	33	50	33	50	33	50	33	50
Cubic Feet of Amendment to be Added	0.3	0.5	1.8	2.7	2.9	4.4	3.6	5.5	8.8	13.0	16.0	24
Container Size	**30 in**		**36 in**		**42 in**		**48 in**		**54 in**		**60 in**	
Amendment (%)	33	50	33	50	33	50	33	50	33	50	33	50
Cubic Feet of Amendment to be Added	33	50	53	80	69	105	103	156	121	184	172	260

turning. High-speed tillers may destroy soil structure. The addition of excessive amendments will lead to improper nutrient incorporation rates. For example, the addition of 3 inches of amendment with a tiller that mixes only 6 inches deep will lead to a 50 percent mixed soil. If a 30 percent addition was intended, only 2 inches of amendment should have been applied and mixed. After tilling, the area should be graded to provide for surface drainage, ensuring that no low spots or pockets remain.

Mulching

Mulches are materials that are spread on the surface of the ground to:

- Retain soil moisture
- Protect against soil erosion
- Protect the roots of plants from heat, cold, or drought
- Keep fruit or plants clean
- Allow earlier planting
- Assist in providing uniform seed germination and plant establishment
- Assist in weed control

Mulch is commonly applied to landscape beds and around trees, shrubs, and bedding plants. Commonly used mulch materials include straw, plastic, wood chips or bark, and fibrous cellulose materials such as those used in hydroseeding. Combinations of these products may also be used. Generally, mulching involves the spreading of materials at depths of ½ to 3 inches (Table 6-7). If applying shredded bark mulch, 3 inches of mulch over the top of the soil is recommended. Rodents find mulch easy to burrow through, so deep mulching may encourage rodent habitation. Rodents may feed on plant roots or girdle stems of shrubs by feeding on plant bark. In areas prone to rodent feeding, keep mulch several inches away from tree trunks.

Top-Dressing

It is not possible to physically incorporate amendments into established turfgrass. However, a light application of organic amendments over the top of a lawn, known as *top-dressing*, allows for the gradual improvement of soils under turf. Amendment top-dressing should be limited to depths of ⅛ to ¼ inch. Top-dressing thicker than ¼ inch on newly seeded lawns or ground covers may interfere with germination and water movement, and may promote fungal

TABLE 6-7
Volume of Materials Required for Top-Dressing or Mulching

	Approximate Volume of Material	
Depth	**per 1,000 sq ft**	**per Acre**
inches	*cubic feet*	*cubic yards*
1/8	10	16
1/4	20	32
3/8	30	48
1/2	40	64
1	80	128
2	160	256
3	320	512

diseases. On newly seeded areas, the use of faster-decomposing materials is recommended. Materials with longer residual duration may contribute to eventual water infiltration problems.

CHEMICAL AMENDMENTS

A highly productive soil is chemically balanced. A chemically balanced soil has a pH near neutral, has adequate levels of all nutritional elements, does not have elevated levels of toxic elements, and has proportions of beneficial base cations that do not result in low water permeability or soil physical problems. A *chemical amendment* is defined as any substance used for the purpose of promoting plant growth or improving the quality of plants by conditioning soils solely through chemical means. The remainder of this chapter will discuss the purpose of using chemical amendments to modify improper soil pH and to correct conditions where excess sodium has led to unmanageable soil physical conditions.

Acidic Soils

Most Western US soils have a neutral to basic (>7.0) pH. However, there are certain conditions, either natural or man-made, that can cause soils to become acidic. Soils with a pH less than 5.5 are considered to be strongly acidic. Acidic soils may be detrimental to plants for several reasons:

- Soluble aluminum and manganese can become toxic to plants
- Microorganisms involved in nitrogen and sulfur transformations are adversely affected
- Phosphorus availability is reduced
- Calcium and magnesium deficiencies occur due to excessive leaching of these nutrients

Acid-tolerant plants, also known as "acid-loving" plants, grow best in soils with a pH between 4.0 and 5.5. Examples of some acid-tolerant ornamental plants are listed in Table 6-8. Even acid-tolerant plants will be damaged by excessively low soil pH

TABLE 6-8
Some Genera of Acid-Tolerant Ornamental Plants

Abelia	*Juniper*	Raspberry
Camellia	*Ligustrum*	*Rhododendron*
Catalpa	*Liquidambar*	*Spirea*
Citrus	*Lonicera*	Strawberry
Dichondra	*Magnolia*	*Salix*
Gardenia	*Nandina*	*Syringa*
Gladiolus	*Persea*	*Verbena*
Hibiscus	*Photinia*	*Vinca*
Hydrangea	*Pyracantha*	*Wisteria*
Iris	*Quercus*	

Handreck, K.A., and N. Black. 2002. *Growing Media for Ornamental Plants and Turf.* UNSW Press, Sydney, Australia.

(pH < 4.0). Correcting soil pH is necessary to optimize plant nutrient availability and plant health. When the proper pH is achieved prior to planting and is maintained during the growing season, many nutritional concerns will be eliminated.

The optimum pH range for soilless media is normally 0.5 to 1.0 pH units lower than for mineral soils. This is because nutrient availability, especially for micronutrients, is normally optimized at a slightly lower pH in media. Most species and cultivars of horticultural plants can be grown in organic potting media with pH values in the range of 5.5 to 6.3 when the media has balanced nutrient availability for all macro- and micronutrients. More information on soilless media can be found in Chapter 4.

Acid soils are generally found where rainfall is high, where repeated applications of acid-forming fertilizers have been used on poorly buffered soils, or where peat or organic deposits occur. Landscape grading often exposes soils with poor buffering capacity that may be prone to acidification. Acid soils are more readily formed from weathering of low pH igneous rocks, such as granite, and sedimentary rocks, such as sandstone. The weathering of basic igneous rocks, such as basalt, has a minimal impact on soil pH.

Soils become acidic when the base cations on the soil colloids, primarily calcium, are replaced by hydrogen ions. Ion exchange and other adsorptive reactions associated with the colloids are important to the rate of acidification. The type of clay and the amount of organic matter influences these reactions.

The nature of soil acidity is complex. One part, called the *active acidity,* is made up of the hydrogen ions in the soil solution. These are the ions measured when the pH of a soil is determined. Another component of soil acidity is the *potential acidity,* representing hydrogen and acid-forming ions held or bound on colloidal surfaces. The combination of potential and active acidity is called *total acidity*. Potential acidity is usually much greater than active acidity. Since clay particles and organic matter have large adsorptive surface areas, soils high in clay or organic matter usually have a much higher total acidity than sandy soils. It takes more lime to raise the pH of soils containing large amounts of clay or organic matter than for a sandy soil.

Liming Acid Soils

One of the most common uses of chemical amendments is to modify low soil pH. Amending soils with lime can ameliorate many or all of the problems associated with low soil pH. Table 6-9 shows the effect

TABLE 6-9

Amount of Limestone Needed to Change the Soil Reaction (Approximate)

	Lime Requirements, lb/1,000 sq ft[1]	
Soil Texture	**pH 4.5 to 5.5**	**pH 5.5 to 6.5**
Sand and loamy sand	23	28
Sandy loam	37	60
Loam	55	78
Silt loam	69	92
Clay loam	87	106
Muck	174	197

[1]Incorporated to a depth of 6 inches.

of finely ground limestone on the pH of soils of different textures and includes a highly organic soil (muck).

To correct soil acidity, materials commonly used for increasing the soil pH are the carbonates, oxides, hydroxides, and silicates of calcium and magnesium. One should never apply limestone or dolomite to soils that have a pH of 7.0 or greater. The solubility of calcium carbonate decreases logarithmically as pH increases from very acid to near neutral pH. As pH approaches and goes above 7.0, calcium carbonate is essentially insoluble.

Table 6-10 lists the liming materials most commonly used to treat acid soils. The liming materials in the table are given in terms of percent calcium carbonate equivalent. Hydrated and burned lime has been heated to drive off carbon dioxide found in lime, which results in calcium carbonate equivalent values that can be greater than 100 percent. Additional liming materials include potassium bicarbonate, which when dissolved in water, has been used in potting mixes to raise the pH of the media quickly. Calcium-based limestone is termed *calcitic limestone*. Limestone that contains between 10 to 50 percent magnesium carbonate is called *dolomitic limestone*. The minimum percentage of magnesium required to be called a dolomitic limestone is defined by fertilizer regulatory agencies on a state-by-state basis. The acid-neutralizing

TABLE 6-10

Common Liming Materials

Name	Chemical Formula	% $CaCO_3$ Equivalent	Source
Shell meal	$CaCO_3$	95	Natural shell deposits
Limestone	$CaCO_3$	100	Pure form, finely ground
Hydrated lime	$Ca(OH)_2$	120–135	Steam burned
Burned lime	CaO	150–175	Kiln burned
Dolomite	$CaCO_3 \cdot MgCO_3$	110	Natural deposits
Sugar beet lime	$CaCO_3$	80–90	Sugar beet by-products
Calcium silicate	$CaSiO_3$	60–80	Slag
Power plant ash	CaO, K_2O	5–50	Wood-fired power plants
Cement kiln dust	CaO, $CaSiO_3$	40–60	Cement plants

value of a commercial liming material will depend on its purity and particle size. A material does not qualify as a liming compound just because it contains calcium or magnesium. Gypsum has little direct effect on soil pH. However, soluble calcium provided by gypsum can exchange with hydrogen adsorbed to clay and organic matter, thereby raising acid soil and media pH to no more than 6.3.

The acid-neutralizing reaction with limestone can be written:

$$CaCO_3 + 2H^+ \rightarrow Ca^{2+} + H_2O + CO_2\,(g)$$

The value of a liming material is governed by its mineral composition, purity, and the degree of fineness. Some established standards indicate that all material must pass through a 60-mesh screen to have a full-efficiency rating. Liming materials finer than 100 mesh will react more rapidly but are more difficult to apply due to their dusty nature.

There are broad benefits to liming acid soils. Calcium and magnesium are essential plant nutrients and their addition as amendments may provide direct nutrient value. Additionally, the correction of adverse chemical, physical, and biological conditions may result in striking improvements in plant growth. Phosphorus availability is greatest at a soil pH of 6.5 to 7.0. Toxicities of certain elements are reduced when soils have a pH greater than 6.0. Beneficial microbial activity is favored at a neutral or near-neutral pH. Microbiological processes such as nitrification, nitrogen fixation, and decomposition of plant residues are optimized at neutral soil pH. At the same time, soil aggregation and structural development may be improved.

Pre-plant applications of lime are the best approach to preventing problems that may occur later. If exchangeable magnesium is high relative to calcium in the irrigation water or on the soil colloids, it may be best to use calcitic limestone. If exchangeable calcium is high relative to magnesium, the use of dolomitic lime is preferred. Dolomitic lime sources are sometimes slower to dissolve and react than calcitic lime sources.

If the initial applications of lime do not maintain the desired pH, additional applications may be needed during the growing season. Slow increases in pH can be achieved by spreading lime onto the soil surface followed by irrigation. Finely ground flowable lime added to very pure, low salt irrigation water can have a rapid effect on increasing pH. However, water with high alkalinity (bicarbonate and carbonate greater than 1.5 meq/L) may react with lime added to irrigation water and could plug microirrigation emitters. Limited amounts of potassium carbonate or potassium hydroxide have been used in both soils and soilless media for short term increases in pH, but this practice must be done with care. Nitrate forms of nitrogen will also cause a slow increase in soil pH. One should limit the application of ammoniacal sources of nitrogen to highly acidic soils.

Acidifying Soils

Many acid-loving plants require a low soil pH. Plants in the families of Ericaceae (azalea, rhododendron, heather, blueberry) and Theaceae (*Ternstroemia* spp., camellia) require an acidic soil pH (<5.5). The pH can be reduced in several ways. Physical amendments, such as peat, will reduce pH. Acidifying chemical amendments include the use of aluminum sulfate, iron sulfate, and elemental sulfur. All of these sources react in the soil to form sulfuric acid, which dissociates to lower pH. The following reaction equation is an example of how aluminum sulfate reacts with water to form sulfuric acid.

$$Al_2(SO_4)_3 + 3H_2O \rightarrow Al_2O_3 + 3H_2SO_4$$

All of these materials can be mixed with the soil as preplant amendments. The sulfur will take several weeks or months to become oxidized to sulfuric acid depending on the particle size, temperature, and soil moisture. Elemental sulfur is converted in the soil to sulfuric acid by bacteria in the genus *Thiobacillus*. For information on proper rates of elemental sulfur (S) to apply to acidify soil, see Table 6-11. For a quicker response, iron sulfate may be preferred, since it avoids any potential problems associated with aluminum. Both iron sulfate and aluminum sulfate are water-soluble. They can be dissolved in irrigation water and applied by fertigation. Some nurseries routinely acidify all the irrigation water used for plant production. Instead of using acid-forming amendments, sulfuric acid, and urea sulfuric acid can be applied to soil or irrigation water directly. Sulfuric acid is extremely dangerous and should only be handled by those who are qualified and

TABLE 6-11

Pounds of Elemental Sulfur Needed to Lower a Silt Loam Soil[1] pH to a Depth of 6 Inches (Approximate)

Present	Desired Soil pH				
pH	6.5	6.0	5.5	5.0	4.5
	------- lb S per 100 sq ft -------				
8.0	3.0	4.0	5.5	7.0	8.0
7.5	2.0	3.5	4.5	6.0	7.0
7.0	1.0	2.0	3.5	5.0	6.0
6.5	—	1.0	2.5	4.0	4.5
6.0	—	—	1.0	2.5	3.5

[1]For sandy soils, reduce amount by 1/3; for clayey soils, increase amount by 1/2; if aluminum sulfate is used, multiply by 6.9.

Mitchell Jr., C.C., and J.F. Adams. 2000. *Soil Acidity and Liming-Part 2.* http://hubcap.clemson.edu/~blpprt/lowerpH.html

trained in its use. A safer acid is produced when urea is reacted with sulfuric acid to make urea sulfuric acid.

All ammonium-based fertilizers, including urea, will have an acidifying effect on media after being oxidized to nitrate by microbial activity. In addition, plant uptake of ammonium will result in acidification through the release of H^+ ions by roots into the soil or media solution. Over-fertilization with ammonium-based fertilizers is not recommended as a technique to reduce pH since ammonium toxicity may result. Application of nitrogen fertilizer in excess of plant uptake may result in pollution of ground and surface waters. Ammonium and urea fertilizers produce acidity through the following simplified reaction:

$$NH_4^+ + 2O_2 \xrightarrow[\text{bacteria}]{\text{nitrifying}} NO_3^- + H_2O + 2H^+$$

Sodic Soils

Sodic soils contain excessive amounts of exchangeable sodium and are often referred to as *alkali* soils since sodium is an alkali earth metal. This term can lead to confusion with *alkaline* soils, which are soils with a pH greater than 7.0, so the term “sodic” is preferred. Sodic soil conditions increase soil pH, deteriorate soil structure, cause surface crusting, and result in poor drainage. A soil test can determine the severity of sodium build-up, and whether amendments are necessary to reduce exchangeable sodium. For more information on soil testing for sodium, refer to Chapter 11.

Irrigating with water containing a high proportion of sodium or having soils with sodium salts will result in sodium replacing calcium and magnesium on the cation adsorption sites of the clay or organic matter in soil. In addition to natural sources, sodium concentrations in irrigation water may be elevated if the water is derived from municipal sources containing residues from laundry detergents, water softeners, or pool water. Soil sodium concentrations may be elevated by some fertilizers that contain sodium.

Amending Sodic Soils

Reclamation treatments for sodic soils vary depending on the presence or absence of native calcium carbonate, also referred to as *free lime*. Gypsum, lime-sulfur (calcium polysulfide), or other soluble calcium salts must be applied to replace exchangeable sodium on soils with little or no free lime. Gypsum is most commonly surface applied, but drip-grade gypsum can also be injected into the irrigation water using a suspension injection machine. Soluble calcium displaces sodium from the exchange sites and the sodium is leached from the soil with irrigation water.

The presence of free lime in the soil allows the widest selection of amendments. To test for free lime, place a few drops of dilute acid or vinegar on the soil. Bubbling or fizzing indicates the presence of carbonates or free lime. A quantitative determination of lime content requires a laboratory analysis. If the soil contains free lime, any of the amendments listed in Table 6-12 may be used.

One of the most cost-effective methods of reclaiming a sodic soil containing free lime is to apply elemental sulfur (S). As sulfur is oxidized to form sulfuric acid, free calcium carbonate will be dissolved forming soluble gypsum. Soluble calcium replaces sodium on the exchange sites and brings about a better physical condition, allowing leaching of sodium and excess salts. It should be noted that the reaction of sulfur in the soil is slow and may take one to several years to come to completion. Like other chemical amendments, the particle size of sulfur affects the rate of reaction with smaller particle sizes reacting faster than larger sizes.

TABLE 6-12

Amendments for Sodic Soils and Their Equivalent Amendment Values

		Tons of Amendment Equivalent to:	
Material (100% Basis)	**Chemical Formula**	**1 Ton Pure Gypsum**	**1 Ton Soil Sulfur**
Gypsum	$CaSO_4 \cdot 2H_2O$	1.00	5.38
Soil sulfur	S	0.19	1.00
Sulfuric acid (98%)	H_2SO_4	0.61	3.20
Ferric sulfate	$Fe_2(SO_4)_3 \cdot 18H_2O$	1.09	5.85
Lime sulfur (22% S)	CaS_x	0.68	3.65
Aluminum sulfate	$Al_2(SO_4)_3 \cdot 18H_2O$	1.29	6.94
Ammonium polysulfide	$(NH_4)_2S_x$	0.37	1.95

Applying organic amendments, such as manure, plant residues, peat, sawdust, compost, or bark may be helpful when reclaiming sodic soils. Organic amendments increase soil porosity and facilitate drainage so excess sodium can be leached easier.

Amendment Purity

Recommendations for soil amendment additions are generally made with the assumption that the amendments are 100 percent pure. The values given in Table 6-12 are for 100 percent pure amendments. In most cases, an amendment is not pure. The purity of the amendment material can be found on the product label. The following calculation will determine the amount of an impure material needed to be equivalent to one ton of pure material:

$$\frac{100}{\%\ \text{purity}} = \text{tons of amendment}$$

Example: If a gypsum amendment is 60 percent pure, the calculation would be 100/60 = 1.67 tons of 60 percent gypsum amendment to be equivalent to 1 ton of 100 percent pure gypsum.

Amendment Reactions

The acid-forming materials, such as elemental sulfur, go through three steps. The first step is oxidation with the help of *Thiobacillus* bacteria to form sulfuric acid. The soil must be moist and warm (above 45°F) for this biochemical reaction to occur:

$$\underset{\text{sulfur}}{2S} + \underset{\text{oxygen}}{3O_2} + \underset{\text{water}}{2H_2O} \xrightarrow[\text{bacteria}]{\textit{Thiobacillus sp}} \underset{\text{sulfuric acid}}{2H_2SO_4}$$

Sulfuric acid, either created by sulfur and bacteria or applied directly, reacts with free lime to form calcium sulfate (gypsum), carbon dioxide, and water:

$$\underset{\text{sulfuric acid}}{H_2SO_4} + \underset{\text{limestone}}{CaCO_3} \longrightarrow \underset{\text{calcium sulfate}}{CaSO_4} + \underset{\text{carbon dioxide}}{CO_2\,(g)} + \underset{\text{water}}{H_2O}$$

Soluble calcium in calcium sulfate, from either reactions of sulfuric acid and free lime or applied directly, exchanges with exchangeable sodium as such:

$$\underset{\text{calcium sulfate}}{CaSO_4} + \underset{\text{exchangeable sodium}}{Na\text{-}CEC} \longrightarrow \underset{\text{exchangeable calcium}}{Ca\text{-}CEC} + \underset{\text{sodium sulfate}}{Na_2SO_4}$$

Managing Sodic Soils

It may be impractical to reclaim sodic soils completely or to maintain these soils in a low-sodic condition. The reasons may include the high cost of reclamation, the inability to drain the soils adequately, or lack of good quality water to leach excess sodium. The costs of amendments and their application must be considered. For more information on saline soils refer to Chapter 5. Practices that aid in the management of sodic and saline soils include the following:

- Select plant varieties that are tolerant of salts and/or sodium
- Use planting procedures that minimize salt accumulation in the root zone
- Use of site preparation procedures that provide a low-salt environment

- Use of irrigation water to maintain high soil water content to dilute or leach salts away from the root zone
- Use of physical or organic amendments to improve soil structure
- Improve internal drainage with deep tillage to break up hardpans
- Use of chemical amendments

Before a satisfactory management program can be developed, it is essential to know the soil's physical and chemical characteristics, the quality and quantity of irrigation water used, the climate at the growing site, and the economics of the situation. Consulting with a soil scientist or Certified Crop Adviser and having appropriate soil and water tests conducted are essential steps in soil reclamation and in the management of soil problems.

SUPPLEMENTARY READING

American Society of Agricultural Engineers. 1985. Drip/trickle irrigation in action. Proceedings of the third international drip/trickle congress. Vol. 1 and 2.

Ayers, R.S., and D.W. Westcot. 1985. *Water Quality for Agriculture.* Food and Agricultural Organization (FAO) of the United Nations. FAO Irrigation and Drainage Paper 29.

Branson, R.L. 1979. Soil Amendments: What Can They Really Do? *Western Landscape News.*

Chabra, R. 1996. *Soil Salinity and Water Quality.* A.A. Balkema Publisher, Brooksfield, Vermont.

Grattan, S.R. 2006. *Irrigation Water Salinity and Crop Production.* Division of Agriculture and Natural Resources University of California. Publication 8066.

Handreck, K., and N. Black. 2002. *Growing Media for Ornamental Plants and Turf,* 3rd ed. University of New South Wales Press Ltd., Sydney, Australia.

Hanson, B., S.R. Grattan, and A. Fulton. 1999. *Agricultural Salinity and Drainage*. Division of Agriculture and Natural Resources University of California. Publication 3375.

Horneck, D.A., J.W. Ellsworth, D.M. Sullivan, and R.G. Stevens. 2007. *Managing Salt-Affected Soils*. Pacific Northwest Extension PNW 601-E.

Kamprath, E.J. 1991. *Soil Acidity and Liming*. Agronomy Monograph No. 12. American Society of Agronomy.

Lascano, R.J., and R.E. Sojka eds. 2007. *Irrigation of Agricultural Crops*, 2nd ed. Agronomy Monograph No. 30. American Society of Agronomy.

McNeely, W.H., and W.C. Morgan. 1968. Review of soil amendments. *Turfgrass Times*.

Morgan, W.C., J. Letey, S.J. Richards, and N. Valoras. 1968. The use of physical soil amendments in turfgrass management. *California Turfgrass Culture*.

Richards, L.A. ed. 1954. Diagnosis and improvement of saline and alkali soils, www.ars.usda.gov/Services/docs.htm?docid=10158&page=2. Accessed July 2012.

Shainberg, I., and J. Shalhevet, ed. 1984. *Soil Salinity Under Irrigation*. Springer-Verlag.

Tanji, K.K. ed. 1990. *Agricultural Salinity Assessment and Management*. ASCE Manuals and Reports on Engineering Practice. #71.

Tisdale, S.L., W.L. Nelson, J.D. Beaton, and J.L. Havlin. 1993. *Soil Fertility and Fertilizers*. 5th edition. The Macmillan Company.

Warneke, J.E., and S.J. Richards. 1974. *Evaluating Soil Amendments*. California Agriculture.

Westerman, R.L., ed. 1990. *Soil Testing and Plant Analysis*. Soil Science Society of America.

CHAPTER 7

Fertilizers

Soil serves as a storehouse for plant nutrients and normally provides a substantial amount of a plant's nutrient requirements. Few soils contain all the essential nutrients in the proper balance to allow plants to reach their full growth potential. Soilless media supplies, without added fertilizer, only limited plant nutrition. Under most conditions, plant growth and quality can be improved by an application of supplemental nutrients. A *fertilizer* is any material containing one or more of the essential plant nutrients that is added to the soil, soilless media, or applied to plant foliage for the purpose of supplementing the plant nutrient supply.

The earliest fertilizer materials were animal manures, plant and animal residues, ground bones, and potash salts derived from wood ashes. Three major developments in the 19th century in Europe were the forerunners of the modern fertilizer industry:

- 1839—The discovery of potassium salt deposits in the German states
- 1842—The treatment of ground phosphate rock with sulfuric acid to form superphosphate
- 1910—The development of the theoretical principles for combining hydrogen and atmospheric nitrogen to form ammonia

Researchers and practitioners continuously work together to improve the availability of plant nutrients. This includes finding renewable sources of nutrients, improving fertilizer application methods, and enhancing techniques that maintain and improve soil quality and plant health. The horticultural industry uses laboratory and field testing to identify the specific forms and amounts of nutrients a plant needs, formulate or acquire a fertilizer that meets plant needs, and then apply the fertilizer at the appropriate time in a plant's lifecycle for optimum uptake.

TYPES OF FERTILIZERS

Based upon their primary nutrient content (N, P_2O_5, K_2O), fertilizers are designated as single-nutrient or multi-nutrient fertilizers. It is important to note that fertilizer percentages of phosphorus and potassium in the United States are not reported in terms of their elemental concentrations. Instead, fertilizers containing phosphorus are reported as *phosphorus pentoxide* or P_2O_5 (commonly referred to as "phosphate"). Likewise fertilizers containing potassium are reported in terms of *potassium oxide* or K_2O (often referred to as "potash"). These designations were developed from older chemical analysis methods for determining fertilizer grades that produced oxide forms of each element. Refer to the section later in this chapter titled "Nutrient Conversion Factors" (page 201) and Table 7-10 (page 203) for more information on conversions between elemental and oxide forms of phosphorus and potassium.

Single-nutrient fertilizers are sometimes referred to as *simple fertilizers.* Multi-nutrient fertilizers may be referred to as *blended fertilizers* or *homogenous fertilizers.* Fertilizers containing one or more primary nutrients are given a numerical designation consisting of three numbers. This three-number designation is called a *grade* and represents, respectively, the weight percent of nitrogen (N), phosphate (P_2O_5), and potash (K_2O) contained in the fertilizer. The *fertilizer ratio* is the relative proportion of each of the primary nutrients. For example, a 12-12-12 grade fertilizer has a 1:1:1 ratio, and a 21-7-14 grade fertilizer has a 3:1:2 ratio. A zero in either a grade or a ratio designation indicates that a particular nutrient is not included in the fertilizer. For example, the grade designation for diammonium phosphate is 18-46-0 indicating that potash is not present in this fertilizer.

The particle size of dry fertilizer products is characterized by two parameters developed by the Canadian Fertilizer Institute—the Size Guide Number (SGN) and the Uniformity Index (UI). *Size Guide Number* is defined as the median particle size in millimeters times 100. A fertilizer with a large SGN will have a greater particle size than one with a smaller SGN. For example, an agricultural granular fertilizer with a 270 SGN would have a median particle size of 2.7 mm.

The SGN of a fertilizer is important for both blending and application. A dry fertilizer blend consisting of widely varying particle sizes is likely to separate during transportation and should be avoided. Special applications, such as on golf greens and fairways, require small particle size fertilizers. Size ranges for granular fertilizers are from 200 to 320 SGN. Mini size fertilizers for turf application range from 120 to 150 SGN. Greens grade fertilizers range from 90 to 100 SGN.

The *Uniformity Index* (UI) is the particle size at which 95 percent of the material is retained, divided by the particle size at which 10 percent of the material is retained, multiplied by 100. The closer the Uniformity Index is to 100, the closer the particles are to the same size. Most dry fertilizers have a UI between 20 and 50.

NITROGEN FERTILIZERS

Anhydrous Ammonia (82-0-0)

Atmospheric nitrogen is the source of all nitrogen used by plants. Nitrogen gas comprises about 78 percent of the earth's atmosphere. However, with the exception of legumes, plants cannot utilize this inert gas directly. Atmospheric nitrogen is chemically fixed to form ammonia, the principal source for all nitrogen-containing fertilizers (Figure 7-1). The production of ammonia involves the energy-intensive reaction of nitrogen and hydrogen gasses known as the Haber-Bosch process. Natural gas, or methane, is normally the source of hydrogen and energy for the reaction. Hydrogen and nitrogen combine in the presence of an enriched iron oxide catalyst at temperatures ranging from 570°F to 1,000°F and pressures ranging

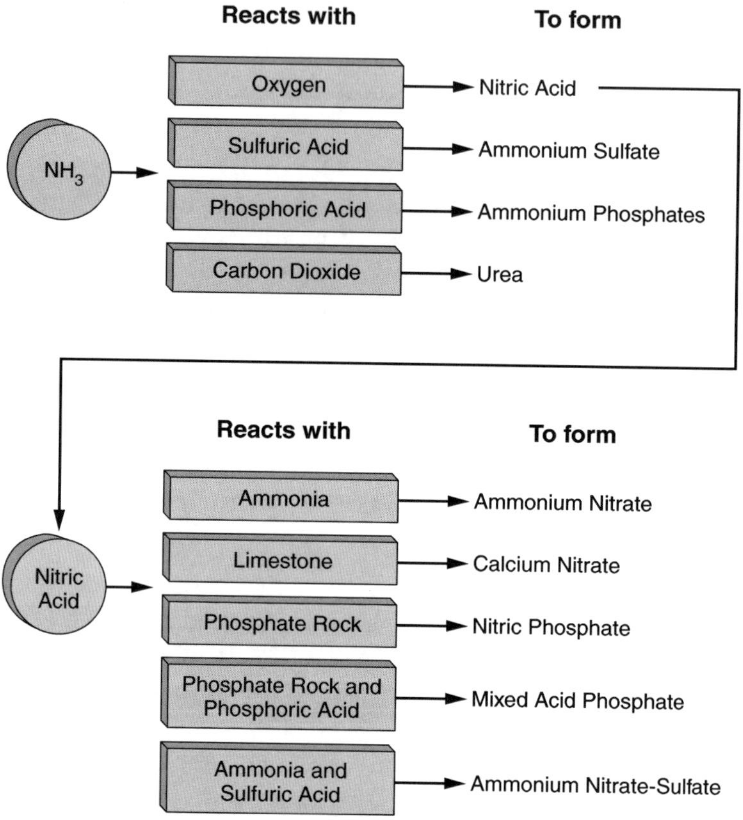

FIGURE 7-1. Conversions of ammonia to various nitrogen fertilizers.

from 2,000 to 3,500 pounds per square inch to form ammonia, according to the following equation:

$$\underset{\text{hydrogen}}{3H_2} + \underset{\text{nitrogen}}{N_2} \xrightarrow[\text{Iron oxide catalyst}]{\text{High temperature \& pressure}} \underset{\text{ammonia gas}}{2NH_3}$$

Gaseous ammonia is lighter than air, readily liquefies when compressed, and is about 60 percent heavy as water. Ammonia is readily absorbed in water up to concentrations of 30 to 40 percent by weight. The high vapor pressure of anhydrous ammonia at ordinary temperatures requires that it be transported in pressurized

Properties of Anhydrous Ammonia

Nitrogen content	82%
Color	Colorless
Odor	Pungent, sharp
Molecular weight	17 (g/mole)
lb per gal @ 60°F	5.14
lb per cu ft @ 60°F	38.45
Boiling point at 1 atmosphere pressure	–28.03°F
Freezing point at 1 atmosphere pressure	–107.86°F
Calcium carbonate equivalent (lb $CaCO_3$/100 lb fertilizer)	148

containers, generally with a minimum working pressure of 265 pounds per square inch.

Although the supply of nitrogen from the air is virtually infinite, sources of hydrogen are limited. In the United States, almost all modern ammonia production facilities use natural gas as the hydrogen source. One ton of ammonia requires about 33,500 cubic feet of natural gas to supply the hydrogen required. Alternative sources such as naphtha, a hydrogen-rich hydrocarbon refined from petroleum, are frequently used in foreign production facilities.

Anhydrous ammonia has traditionally been a relatively inexpensive source of nitrogen fertilizer. Because of safety, handling, and environmental concerns, anhydrous ammonia use in North America continues to decline.

Aqua Ammonia (20-0-0)

Anhydrous ammonia dissolved in water forms *aqua ammonia*, commonly called "aqua". Fertilizer grade aqua contains 20 percent nitrogen, all in the ammonium form. Under normal temperatures, aqua ammonia has some free ammonia, but the vapor pressure is low. This makes it possible to store this liquid fertilizer in low-pressure tanks and to apply it with less-expensive equipment than is required for anhydrous ammonia.

To minimize the loss of nitrogen, aqua ammonia, like anhydrous, should be injected below a soil or water surface.

Direct soil applications of aqua need not be injected as deeply as those of anhydrous ammonia. Neither anhydrous ammonia nor aqua should be applied through low pressure micro-irrigation systems. These materials raise water pH upon injection, resulting in precipitation of calcium and magnesium carbonates that may plug emitter flow paths. Volatilization losses of ammonia may also occur.

Ammonium Nitrate (34-0-0)

Ammonium nitrate was not used extensively as a fertilizer until after World War II. It is manufactured by reacting nitric acid with anhydrous ammonia (Figure 7-2).

Nitric acid is produced by the oxidation of NH_3 with air in the presence of a catalyst, usually platinum. The initial oxidation reactions:

$$\underset{\text{ammonia}}{4NH_3} + \underset{\text{oxygen}}{5O_2} \xrightarrow{\text{catalyst}} \underset{\text{nitrous oxide}}{4NO} + \underset{\text{water}}{6H_2O}$$

$$\underset{\text{nitrous oxide}}{2NO} + \underset{\text{oxygen}}{O_2} \longrightarrow \underset{\text{nitric oxide}}{2NO_2}$$

The NO_2 is then absorbed in water to form nitric acid.

$$\underset{\text{nitric oxide}}{3NO_2} + \underset{\text{water}}{H_2O} \longrightarrow \underset{\text{nitric acid}}{2HNO_3} + \underset{\text{nitrous oxide}}{NO}$$

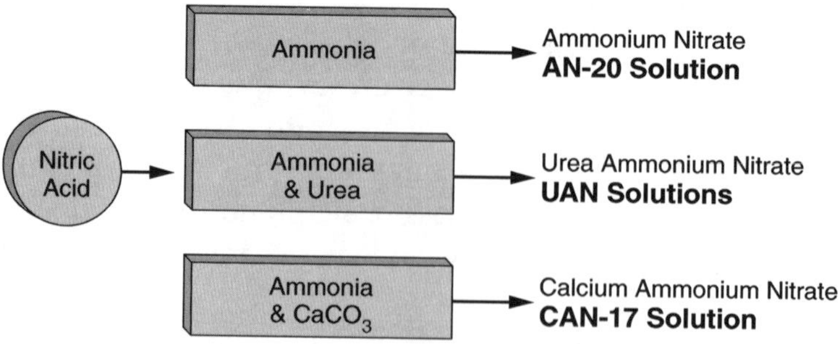

FIGURE 7-2. Liquid nitrogen solutions.

Nitric acid and ammonia react to form ammonium nitrate.

$$\underset{\text{nitric acid}}{HNO_3} + \underset{\text{ammonia}}{NH_3} \longrightarrow \underset{\text{ammonium nitrate}}{NH_4NO_3}$$

The final product is concentrated, prilled, or granulated and then coated to prevent caking. Commercial ammonium nitrate contains about 34 percent nitrogen. Dry ammonium nitrate is a strong oxidizer and can be explosive. Security concerns have restricted availability of dry ammonium nitrate. Safer alternatives, such as liquid ammonium nitrate solution (AN-20) or dry calcium ammonium nitrate (CAN-27), have replaced the commercial use of dry ammonium nitrate.

Properties of Ammonium Nitrate

Nitrogen content	34%
Color	White
Molecular weight	80 (g/mole)
lb per cu ft	45–62
Angle of repose	37° – 40°
Critical relative humidity @ 68°F, 20°C	63.3%
Calcium carbonate equivalent (lb $CaCO_3$/100 lb fertilizer)	60

Ammonium Sulfate (21-0-0-24S)

Ammonium sulfate is a widely used fertilizer in the western United States. It contains 21 percent nitrogen and 24 percent sulfur and is one of the oldest forms of solid nitrogen fertilizer.

Manufactured ammonium sulfate is made in a neutralizer-crystallizer unit by reacting anhydrous ammonia with sulfuric acid.

The reaction is as follows:

$$\underset{\text{ammonia}}{2NH_3} + \underset{\text{sulfuric acid}}{H_2SO_4} \longrightarrow \underset{\text{ammonium sulfate}}{(NH_4)_2SO_4}$$

Most fertilizer-grade ammonium sulfate is produced by direct crystallization. Substantial quantities are also available as a by-product from various industrial processes.

Properties of Ammonium Sulfate

Nitrogen content	21%
Sulfur content	24%
Color	White
Molecular weight	132 (g/mole)
lb per cu ft	66 – 68
Angle of repose	28° – 30°
Critical relative humidity @ 68°F, 20°C	81%
Calcium carbonate equivalent (lb $CaCO_3$/100 lb fertilizer)	110

Calcium Nitrate (15.5-0-0-19Ca)

The commercial grade of calcium nitrate—chemically known as ammonium calcium nitrate decahydrate—is a white, hygroscopic, dry, water-soluble product containing 15.5 percent nitrogen and 19 percent calcium. It is a co-product of the nitric phosphate process where nitric acid is reacted with crushed phosphate ore and neutralized with anhydrous ammonia. The general reaction is:

$$Ca^{2+} + HNO_3 \xrightarrow{NH_3} 5Ca(NO_3)_2 \cdot NH_4NO_3 \cdot 10H_2O + \text{Nitric Phosphate Fertilizers}$$

Ca^{2+}: calcium from phosphate ore; HNO_3: excess nitric acid; NH_3: ammonia; $5Ca(NO_3)_2 \cdot NH_4NO_3 \cdot 10H_2O$: ammonium calcium nitrate decahydrate

The ammonium calcium nitrate decahydrate is separated from the mother liquor and concentrated via successive evaporation and filtration. The final product is granulated and coated to reduce moisture absorption during storage, handling, and application.

Sodium Nitrate (16-0-0)

This dry nitrogen material contains 16 percent nitrogen. A very small amount is manufactured in the United States, but a considerable quantity of this mined, natural product is imported from Chile. It is used primarily in the mid-South and the Southeast as a supplemental nitrogen source on specialty plants. Mined sodium nitrate carries a restricted certification as a fertilizer for organic farming, but the quantity used for this purpose is limited.

Urea (46-0-0)

Urea has one of the highest nitrogen percentages of all solid fertilizers. In production, ammonia and carbon dioxide are reacted in a special vessel at temperatures between 340°F to 410°F and pressures ranging from 2,500 to 6,000 pounds per square inch. The reactions are:

$$\underset{\text{ammonia}}{2NH_3} + \underset{\text{carbon dioxide}}{CO_2} + \underset{\text{water}}{H_2O} \longrightarrow \underset{\text{ammonium carbonate}}{(NH_4)_2CO_3}$$

$$\underset{\text{ammonium carbonate}}{(NH_4)_2CO_3} \longrightarrow \underset{\text{urea}}{(NH_2)_2CO} + \underset{\text{water}}{2H_2O}$$

The concentrated liquid from these reactions contains about 80 percent urea. This may be used directly in urea solutions or further concentrated then prilled or granulated to make solid urea. Dry fertilizer-grade urea contains 46 percent nitrogen.

Urea is highly water-soluble and is less corrosive to equipment than many other fertilizer products. It is a common nitrogen source in dry bulk blends but is incompatible with some fertilizers, particularly those containing even small quantities of ammonium nitrate.

Properties of Urea

Nitrogen content	46%
Color	White
Molecular weight	60 (g/mole)
lb per cu ft	46 – 48
Angle of repose	40°
Critical relative humidity @ 68°F, 20°C	80.7%
Calcium carbonate equivalent (lb $CaCO_3$/100 lb fertilizer)	71

In the past, urea manufacturing processes sometimes resulted in elevated biuret concentrations. *Biuret* is an undesirable by-product formed when molten urea is heated above its melting point, driving off ammonia and forming a molecule with two ureas. In high concentrations, biuret interferes with internal nitrogen metabolism and hinders protein formation in plants. Many soil microorganisms degrade biuret, but the rate is relatively slow. Modern urea manufacturing plants typically produce urea with biuret concentrations of less than 0.9 percent, which does not pose problems for most uses. There are some plant species that are especially sensitive to biuret including ornamental citrus and avocado where toxicity is shown by leaf burning. Seeds are also sensitive to biuret. Urea fertilizer should contain biuret levels of no more than 1.2 percent when soil applied. When urea is applied as a foliar application, it should contain no more than 0.25 percent biuret to prevent phytotoxicity.

UAN-32 (32-0-0)

A liquid blend of urea and ammonium nitrate is a unique fertilizer solution. By combining urea and ammonium nitrate together in solution, one can create a higher analysis liquid than one would get with each individual component dissolved by itself. This permits a stable solution containing more nitrogen than is possible with either single component. This high-analysis urea ammonium nitrate liquid nitrogen solution can reach nitrogen percentages as high as 32 percent and is commonly referred to as UAN-32. UAN-32 has a crystallization temperature of 32°F so locations that have colder temperatures will store UAN-30 or UAN-28, which have crystallization temperatures of 15°F and 1°F, respectively.

Nitrogen Solutions

Nitrogen solutions may contain ammonium, nitrate, and urea nitrogen, the proportion of each being dependent upon the components (Table 7-1). Such liquids are classified as pressure or non-pressure solutions. A pressure solution is one with an appreciable vapor pressure because of the presence of more free ammonia than the solution can hold. Thus, a pressure solution loses nitrogen unless contained in a closed vessel.

Like any solution, nitrogen solutions will exhibit the phenomenon of crystallization or "salting out." *Salting out* is simply the precipitation of the dissolved salts when the temperature drops to the critical point, also known as the crystallization temperature, as indicated in Table 7-1. The crystallization temperature is unique to a given solution and is a function of the chemical properties and the concentration of the components. Higher concentration fertilizer

TABLE 7-1
Composition and Physical Properties of Nitrogen Solutions

	Non-Ammoniated Solutions						Ammoniated Solutions			Aqua Ammonia Solution
	CAN-17	AN-20	UREA-20	UAN-28	UAN-30	UAN-32				
Total N%	17	20	20	28[1]	30[1]	32	37	37	41	20
NH_3%							16.6	15.8	22.2	24.4
NH_4NO_3%	30.0	57.2		39.5	42.2	44.3	66.8	58.5	65.0	
Urea%			43.5	30.5	32.7	35.4		7.7		
Water%	26.0	42.8	56.5	30.0	25.1	20.3	16.6	18.0	12.8	75.6
Nitrate N%	11.6	10.0		7.0	7.4	7.8	11.7	10.2	11.4	
Ammonium N%	5.4	10.0		7.0	7.4	7.8	25.3	23.2	29.6	20.0
Urea N%			20.0	14.0	15.2	16.4		3.6		
Specific gravity at 60°F	1.5	1.26	1.12	1.28	1.3	1.33	1.19	1.17	1.14	0.912
lb/gal at 60°F	12.6	10.5	9.33	10.7	10.8	11.06	9.91	9.75	9.5	7.6
Crystallization temperature °F	25	41	52	1	15	32	56	28	21	–58
lb N/gal at 60°F	2.15	2.1	1.87	2.98	3.25	3.54	3.67	3.61	3.90	1.52
Vapor pressure psi at 100°F	----------(typically non pressure)---------						1	2	10	1

solutions have higher salt-out temperatures. Once a fertilizer crystallizes, it is often difficult to re-dissolve the precipitate without agitation. Crystallized fertilizer can also be difficult to remove from restricted access fertilizer tanks. By knowing crystallization temperatures, one can choose the correct concentration that will prevent salting out at the forecasted minimum temperatures. Refer to Appendix A for a more extensive list of fertilizer solution crystallization temperatures.

CAN-17 (17-0-0-8.8Ca)

Common in the western United States, calcium ammonium nitrate (CAN-17) is a clear solution containing a mixture of calcium nitrate and ammonium nitrate. This liquid fertilizer is produced by reacting calcium-containing minerals, mainly limestone, with nitric acid. The acidic calcium nitrate liquor is neutralized with ammonia and the nitrogen content is adjusted by adding ammonium nitrate. CAN-17 is used as a source of nitrogen and calcium. Each 50 pounds of nitrogen from CAN-17 contains 25 pounds of soluble calcium. CAN-17 is an important cool season fertilizer since two-thirds of the nitrogen is nitrate-nitrogen. CAN-17 is recommended for use in low pH soils due to its high nitrate content. CAN-17 has been well received for use in drip and micro-sprinkler irrigated production areas.

The corrosive characteristics of nitrogen solutions are presented in Table 7-2. Corrosion inhibitors are added to reduce the

Properties of CAN-17

Nitrogen content	17%
Ammoniacal nitrogen	5.4%-5.8%
Nitrate nitrogen	11.2%-11.6%
Calcium content	7.6%-8.8%
Specific gravity @ 68°F, 20°C	1.51
lb per gal @ 68°F, 20°C	12.6
Vapor pressure	Non-pressure
pH	6.2-6.4
Color	Clear or dyed
Salting-out temperature	30°F
Equivalent acidity (lb $CaCO_3$/100 lb N)	53

TABLE 7-2

Corrosive Characteristics of Various Metals, Alloys, and Other Materials by Non-Pressure Nitrogen Solutions

Not Corroded	Corroded	Materials Destroyed Rapidly
Aluminum and Aluminum Alloys types 3003, 3004, 5052, 5154, and 6061	Carbon Steel	Copper
	Cast Iron	Brass
		Bronze
Stainless Steel types 303, 304, 316, 347, and 416		Monel
		Zinc
Rubber		Galvanized metals
Neoprene		Usual die castings
Polyethylene, Polypropylene		Concrete
Vinyl resins; PVC, CPVC		
Glass, fiberglass		
Nylon		
Teflon®		
Viton®		
Kynar®		
Hylar®		

chemical attack of nitrogen solutions on carbon steel, which is often used in construction of tanks and other handling equipment. Corrosion inhibitors commonly used in nitrogen solutions are ammonium thiocyanate (0.1 percent) and borate salts (0.1 to 0.4 percent). Maintaining the solution pH near 7.0 also minimizes corrosion.

PHOSPHORUS FERTILIZERS

The earth's crust in certain areas is richly endowed with natural deposits of fluoroapatite, a phosphate containing mineral. Such deposits are the basic source of all phosphorus fertilizers. These phosphate ores may be of igneous or sedimentary origin, with the latter constituting the bulk of the world's reserves.

The principal world reserves now mined are in North Africa, North America, China, and the former Soviet Union. The important deposits in the United States are located in four areas: Florida; a region made up of Idaho, Montana, Utah, and Wyoming; North Carolina; and Tennessee.

The raw ore has a phosphate (P_2O_5) content of 14 to 35 percent. Lower grade ore must undergo concentration by a process known as beneficiation before it is processed into fertilizer. *Beneficiation* is a process that involves wet screening, hydroseparation, and concentration to remove impurities by flotation of the raw phosphate ore. The resulting product is dried and ground or slurried. At this stage, the P_2O_5 content is 31 to 33 percent, but most of it is still unavailable to plants.

The next step in the process converts the concentrated fluoroapatite ore to more soluble forms of phosphate that can be utilized by plants. Occasionally, finely ground phosphate rock is used directly on acid soils as a source of phosphorus, but its low solubility provides little available phosphorus for plant uptake.

The pathways used by the industry to produce phosphate fertilizers from phosphate rock are shown in Figure 7-3.

Phosphoric Acid and Superphosphoric Acid

Phosphoric acid is an important intermediate in the production of phosphate fertilizers. There are two methods of producing phosphoric acid—the wet process and the furnace-grade method. The wet process is the principal method used by the fertilizer industry. The furnace or thermal process uses electric furnaces operating at extremely high temperatures to burn phosphate rock, thereby removing most impurities. The resulting high phosphate solid material is dissolved in dilute phosphoric acid. This liquid may be further processed to remove heavy metals not eliminated by burning. Furnace-grade acid is costly to produce and is used primarily in food and beverage manufacturing, industrial production, and in formulating highly soluble, specialty fertilizer mixes.

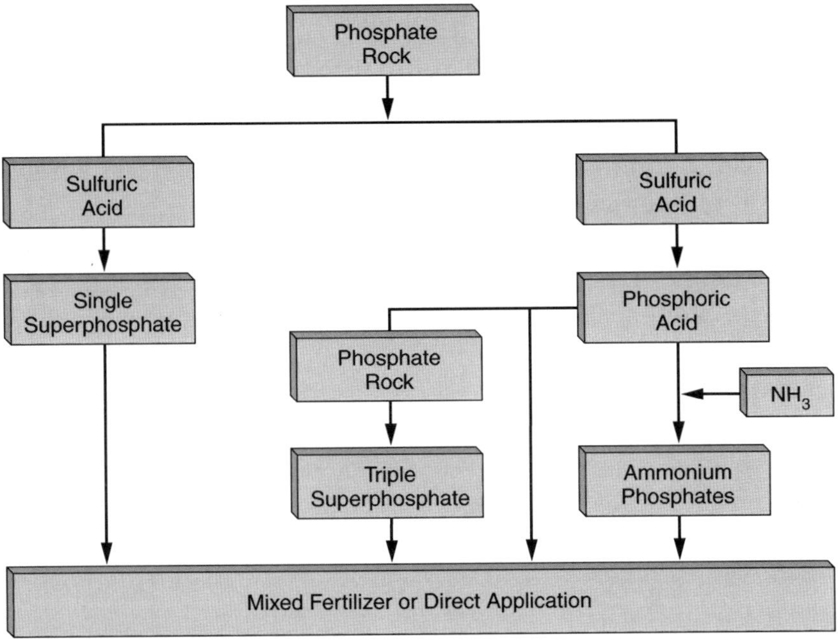

FIGURE 7-3. Pathways for treating phosphate rock in the production of phosphate fertilizers.

Wet-process orthophosphoric acid is produced by reacting sulfuric acid with finely ground phosphate rock (Figure 7-3). The principal chemical reaction in simplified form is:

$$\underset{\text{phosphate rock}}{Ca_{10}F_2(PO_4)_6} + \underset{\text{sulfuric acid}}{10H_2SO_4} + \underset{\text{water}}{20H_2O} \rightarrow \underset{\text{gypsum}}{10CaSO_4{\bullet}2H_2O} + \underset{\text{hydrogen fluoride}}{2HF} + \underset{\text{ortho phosphoric acid}}{6H_3PO_4}$$

Orthophosphoric acid mixed with gypsum is washed with water and filtered to separate the acid from the gypsum. This raw acid contains about 30 percent P_2O_5 and is then concentrated by evaporation to contain between 40 to 54 percent P_2O_5.

Wet-process acid will frequently be green or black as a result of impurities from compounds of Fe, Al, Ca, Mg, F, or organic matter. These impurities may be present in solution or solid form and some may be beneficial as a source of plant nutrients.

Several by-products are formed during the manufacturing of wet-process acid. The principal by-product, impure gypsum, represents a major disposal problem for many fertilizer manufacturers due to the presence of heavy metals.

The trend in the industry is to make and use more concentrated fertilizers. This has prompted the development of superphosphoric acid, a condensation product of orthophosphoric acid. The condensation step is illustrated in Figure 7-4.

The linking of two orthophosphoric acid molecules produces pyrophosphoric acid ($H_4P_2O_7$), the linking of three such molecules gives tripolyphosphoric acid ($H_5P_3O_{10}$), and so on (Figure 7-5).

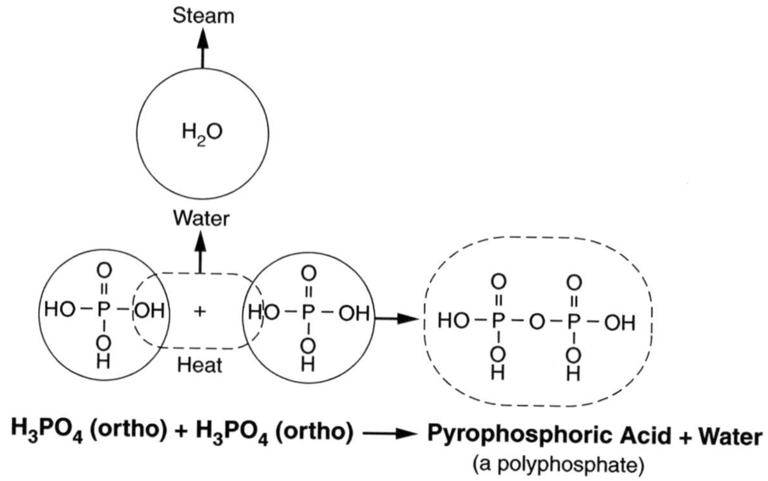

FIGURE 7-4. Condensation removal of water from orthophosphoric acid to produce pyrophosphoric acid.

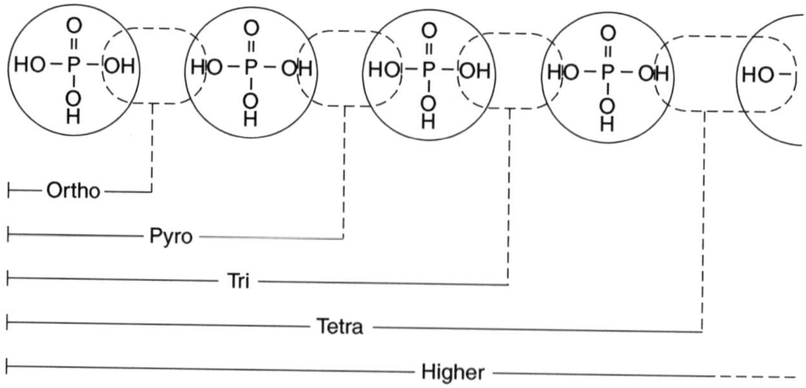

FIGURE 7-5. Linkages of orthophosphoric acid molecules produce various polyphosphoric acids.

Collectively, such acid solutions are called polyphosphoric acids or superphosphoric acids. A *polyphosphoric acid* can be defined as a phosphoric acid whose molecular structure contains more than one atom of phosphorus.

The relationship between the concentration of acid and the type of acid species is shown in Table 7-3.

Single Superphosphate (0-20-0)

Single superphosphate is produced by reacting sulfuric acid with finely ground phosphate rock. The simplified chemical reaction is:

$$Ca_{10}F_2(PO_4)_6 + 7H_2SO_4 + 17H_2O \rightarrow 3Ca(H_2PO_4)_2 \bullet H_2O + 7CaSO_4 \bullet 2H_2O + 2HF$$

phosphate rock	sulfuric acid	water	monocalcium phosphate monohydrate	gypsum	hydrogen fluoride

The highly insoluble phosphate in the phosphate rock is converted to monocalcium phosphate monohydrate in which the phosphate is approximately 85 percent water soluble. Gypsum, another product of this reaction, is intimately mixed with the monocalcium phosphate.

The most common procedure used in the United States combines sulfuric acid and phosphate rock followed by mixing for 1 to 2 minutes. The resulting plastic mass is discharged into a compartment called a den and retained for 1 to 24 hours until the acidulated phosphate solidifies into a hard block. As it is removed from the den, the block is cut into chunks by mechanical excavators equipped with knives. This coarse material is aired for several weeks in storage to allow completion of the chemical reaction. The cured single superphosphate is then pulverized and granulated for shipment and usually contains 20 percent P_2O_5 and 12 percent sulfur.

Triple Superphosphate (0-45-0)

The principal difference between single superphosphate and triple superphosphate is that single superphosphate is reacted with sulfuric acid and triple superphosphate is reacted with phosphoric

TABLE 7-3
Forms of Phosphoric Acid at Various Concentrations[1]

			% of Total Phosphorus as			
Weight % P_2O_5	Equivalent % H_3PO_4	Ortho-phosphate	Pyrophosphate	Tripoly-phosphate	Tetrapoly-phosphate	Higher Poly-phosphate
54.0	75	100				
68.8	95	100				
70.0	97	96	4			
72.0	99	90	10			
75.5	104	53	40	7		
77.0	106	40	47	11	2	
80.0	110	13	35	25	14	13
85.0	117	2	7	8	11	72

[1]This data is for furnace-grade acid. Wet-process acid data is variable with differences occurring between wet-process acid manufacturers.

acid. Finely ground phosphate rock and phosphoric acid in the proper proportions are mixed to produce triple superphosphate. Most superphosphates can be granulated through the addition of water and steam in a rotary drum granulator, followed by drying and screening. The end product contains about 45 percent P_2O_5 and 1 percent sulfur.

The general reaction is:

$$\underset{\text{phosphate rock}}{Ca_{10}F_2(PO_4)_6} + \underset{\text{phosphoric acid}}{14H_3PO_4} + \underset{\text{water}}{10H_2O} \rightarrow \underset{\text{monocalcium phosphate}}{10Ca(H_2PO_4)_2 \bullet H_2O} + \underset{\text{hydrogen fluoride}}{2HF}$$

NITROGEN-PHOSPHATE COMBINATIONS

Ammonium Phosphates

Ammonium phosphates are widely used in bulk blends and for direct application. Common dry ammonium phosphate fertilizers are monoammonium phosphate 11-52-0 (MAP), diammonium phosphate 18-46-0 (DAP), and ammonium phosphate sulfate 16-20-0-15(S). Liquid ammonium phosphate fertilizers are referred to by their analyses and include 8-24-0, 9-30-0, 10-34-0, and 11-37-0. Liquid 8-24-0 and 10-34-0 are produced in the greatest volumes.

Dry ammonium phosphates are produced by reacting ammonia with phosphoric acid in a preneutralization vessel and then further ammoniating the slurry in a rotating ammoniation-granulation unit. Additional nitrogen, phosphate, and sulfuric acid can be introduced into the ammoniation-granulation unit to make the desired grade. The material discharged from the unit is dried, screened, and cooled before being placed in storage. Ammonium phosphate may be present as the diammonium or monoammonium salt or as a mixture of the two.

Solid ammonium polyphosphate fertilizers may be made by ammoniating superphosphoric acid. The acid is ammoniated in a water-cooled reactor at elevated temperature and pressure or in a pipe reactor at high temperature. The anhydrous melt produced is then granulated by mixing it with solid, recycled material in a pugmill, followed by drying and cooling. The analysis of the final product ranges from 11-52-0 to 11-55-0.

Nitric Phosphates

There are three nitric phosphate processes. All three are modifications of the same reaction:

$$\underset{\text{phosphate rock}}{Ca_{10}F_2(PO_4)_6} + \underset{\text{nitric acid}}{20HNO_3} \rightarrow \underset{\text{phosphoric acid}}{6H_3PO_4} + \underset{\text{calcium nitrate}}{10Ca(NO_3)_2} + \underset{\text{hydrogen fluoride}}{2HF}$$

The resulting solution is neutralized with ammonia, concentrated, and granulated. Potassium may be added prior to granulation or prilling. More than 75 percent of the phosphorus is present as highly water soluble monoammonium phosphate and the remaining is relatively insoluble dicalcium phosphate. Modified nitric phosphates are better described as ammonium nitrate phosphates—double salts of ammonium nitrate and ammonium phosphate. The co-product in the process, calcium nitrate liquor, is diverted to neutralization, purification, and granulation to produce a commercial grade of calcium nitrate fertilizer.

POTASSIUM FERTILIZERS

Potassium is widely distributed throughout the world. Extensive deposits of soluble potassium salts are located beneath the earth's surface and in brines of dying lakes and seas. Many of these deposits and brines are sufficiently pure to be mined for plant nutrition and industrial purposes. About 60 percent of the world's potassium reserves are found in North America, predominantly Canada.

Most potassium fertilizers are mined from underground deposits (Figure 7-6). These are recovered by conventional shaft mining and by solution mining in which water is pumped underground to dissolve the ore. The commercial value of a deposit is directly related to its depth, which can vary from a few hundred feet to more than 4,000 feet below the surface. Mining beyond 4,000 feet is presently cost prohibitive. In Saskatchewan, Canada,

FIGURE 7-6. Mining potash ore below the earth's surface.
Source: PotashCorp Inc.

most of the deposits presently mined are more than 3,000 feet below the surface. These deposits tend to be closer to the surface in the north and become deeper as they approach the U.S. border. Deposits in New Brunswick are approximately 2,500 feet below the surface. In Carlsbad, New Mexico, deposits are between 700 and 1,800 feet deep.

Evaporation of water from salt lakes and natural subsurface brines is another source of potassium fertilizers. This method involves precipitating potassium, sodium, and magnesium salts through further concentration from solutions already near the saturation point. This technique is used at the Great Salt Lake and Bonneville Salt Flats in Utah; at Searles Lake, California; and at the Dead Sea in Israel and Jordan.

Properties of the most widely used potassium fertilizers are discussed in the following sections and are presented in Table 7-4.

TABLE 7-4

Properties of Potassium Fertilizers

	Potassium Chloride	Potassium Sulfate	Potassium Nitrate	Potassium Thiosulfate	Potassium Magnesium Sulfate	Mono-potassium Phosphate
Molecular Weight	74.5	174.0	101.0	190.0	415.0	136.0
Specific Gravity	1.98	2.66	2.11	1.46	2.83	2.33
K2O Content, %	63.0	54.0	46.6	25.0	22.7	34.0
Temperature	K_2O concentration in saturated solution, % K_2O[1]					
32°F	13.6	3.7	5.5	25.0	3.8	
50°F	15.0	4.6	8.1	25.0	4.4	19.0[2]
68°F	16.1	5.4	11.2	25.0	5.0	23.0
86°F	17.1	6.2	14.6	25.0	5.6	27.0[2]

[1]All potassium fertilizers are 100% water soluble as defined by the International Association of Official Analytical Chemists.

[2]Estimated.

Potassium Chloride (0-0-60)

A common or commercial term for potassium chloride is muriate of potash (MOP). The term derives from muriatic acid, another name for hydrochloric acid. Fertilizer-grade products range from 60 to 63 percent K_2O depending on purity, and range in color from red to pink to white. The various MOP products are equally effective in supplying potassium to plants. Muriate of potash, like other dry fertilizers, may be purchased in different particle sizes.

Muriate of potash is commonly available in sizes of SGN 285 for granular, 245 for coarse, 85 for standard, and 30 for fine grade. The coarse and granular sizes are best suited for blending with other fertilizer products and for direct application. The fine grade is produced for the liquid fertilizer market and is best suited for solubilizing and injecting into irrigation systems.

Muriate of potash is the least expensive and most widely used potassium fertilizer. It has the disadvantage of containing chloride that can potentially damage sensitive plant materials at high application rates. On the other hand, the chloride content is an advantage in combating a range of plant diseases where chloride is not naturally available.

Potassium Sulfate (0-0-52)

Sulfate of potash (SOP) is the common commercial term for potassium sulfate. Fertilizer grades generally range from 50 to 54 percent K_2O. In addition to potassium, SOP also contains about 17 percent sulfate-sulfur. Most SOP is produced by direct mining or by solar evaporation. Different processes, such as reacting other salts, sulfur, or sulfuric acid with potassium chloride, also produce SOP.

Sulfate of potash is used primarily for those plants that are sensitive to chlorides or in situations where both potassium and sulfur are required. Sulfate of potash has low water solubility, which limits its use to mostly dry applications.

Potassium Magnesium Sulfate (0-0-22-11Mg-22S)

This fertilizer is a double salt of magnesium sulfate and potassium sulfate that is mined from naturally occurring deposits of the mineral langbeinite near Carlsbad, New Mexico. Potassium magnesium sulfate, as its name suggests, supplies potassium, magnesium, and sulfur plant nutrients and is frequently included in dry fertilizer blends to supply these three nutrients. Potassium magnesium sulfate also has the advantage of not containing chloride, making it desirable in situations where chloride sensitive plants are grown.

Potassium Nitrate (13-0-44)

Potassium nitrate is a good source of both potassium and nitrogen. It is widely used as a component of water-soluble fertilizers for injection in irrigation systems and for foliar sprays. Potassium nitrate is less soluble than muriate of potash, but more soluble than sulfate of potash. Potassium nitrate does not contain chloride so it is a potential nutrient source for use on chloride sensitive plants where nitrate-nitrogen is desired. The relatively high cost of production limits its use to medium to high-value plants such as nursery plants.

Potassium Thiosulfate (0-0-25-17S)

Potassium thiosulfate (KTS) is a liquid fertilizer that supplies potassium and sulfur. Potassium thiosulfate is a neutral to basic, chloride-free, clear, liquid solution that can be applied through fertigation alone or in liquid blends with urea solution, urea ammonium nitrate solutions, and 10-34-0. Very dilute potassium thiosulfate can be used as a foliar fertilizer on certain plants, but is known to cause phytotoxicity symptoms under certain conditions and should be used with caution. Potassium thiosulfate is pH sensitive and becomes unstable at pH below 5.5 and above 8.5. The elemental sulfur in thiosulfate will oxidize and generate sulfuric acid that can lower soil and media pH. Greenhouse and nursery managers should use caution when applying potassium thiosulfate to unbuffered soilless media.

Monopotassium Phosphate (0-52-34)

Monopotassium phosphate (MKP) is a specialty fertilizer that is a good source of phosphorus and potassium. It is an acid, chloride-free, dry soluble fertilizer, and is a natural pH buffer with a buffer pH in pure water of 4.5. Monopotassium phosphate can be solubilized and blended with liquid fertilizers that will not salt out at a low pH, and is also safe to use as a foliar fertilizer.

SECONDARY NUTRIENTS

The importance of the major elements nitrogen, phosphorus, and potassium in the fertilizer industry and in horticulture is well established. Secondary nutrients and micronutrients are also essential for plant growth, as discussed in Chapter 2. When guarantees are made for secondary nutrients and micronutrients, the requirement is that they be stated in terms of elemental concentration. Table 7-5 states the contents of secondary nutrients calcium, magnesium, and sulfur in some commonly used fertilizers.

Additional calcium sources for proper plant nutrition are often needed. Soil application of dry or liquid calcium materials and foliar spraying of soluble calcium fertilizers are common practices for various horticultural plants. Non-fertilizer sources that provide additional calcium include lime, gypsum, manure, and irrigation water.

Magnesium sources for crop nutrition include Epsom salts (magnesium sulfate heptahydrate), potassium magnesium sulfate, and magnesium nitrate. Dolomitic lime provides magnesium and calcium when applied to acid soils.

Sulfur for nutrition of plants is provided in many N, P, and K materials (Table 7-5). Additional sulfur–containing fertilizers include:

Source	Percent Sulfur
Elemental sulfur	99
Gypsum	16-18
Sulfuric acid (95%-99%)	32
Ferrous sulfate	11.5
Ferric sulfate	18-19
Calcium polysulfide solution	23
Ammonium polysulfide solution (20% N)	40-45
Ammonium bisulfite solution (8.5% N)	17
Urea–sulfuric acid (28% N)	9
Urea–sulfuric acid (15% N)	16

Other common sources of sulfur include manure, most river water, rainwater, and pesticidal sulfur.

TABLE 7-5

Average Composition of Fertilizer Materials

Fertilizer Materials	Chemical Formula	% N	% P_2O_5	% K_2O	% Ca	% Mg	% S	Equivalent Acidity or Basicity in lb $CaCO_3$[1] Acid	Base
Nitrogen Materials									
Ammonium nitrate	NH_4NO_3	33.5-34.0						100	
Ammonium nitrate sulfate	$NH_4NO_3 \cdot (NH_4)_2SO_4$	30					6.5	68	
Monoammonium phosphate	$NH_4H_2PO_4$	11	52-55					58	
Ammonium phosphate sulfate	$NH_4H_2PO_4 \cdot (NH_4)_2SO_4$	16	20				15	88	
Ammonium phosphate nitrate	$NH_4H_2PO_4 \cdot NH_4NO_3$	23-30	5-23				0-7	65	
Ammonium polysulfide	$(NH_4)_2S_x$	20					40-50		
Diammonium phosphate	$(NH_4)_2HPO_4$	16-18	46-48					70	
Ammonium sulfate	$(NH_4)_2SO_4$	21					24	289	
Ammonium thiosulfate	$(NH_4)_2S_2O_3$	12					26	102	
Anhydrous ammonia	NH_3	82						80	
Aqua ammonia	NH_4OH	20						20	
Calcium ammonium nitrate	$Ca(NO_3)_2PO_4 \cdot NH_4NO_3$	17			7.6-8.8			9	

Calcium nitrate	$5Ca(NO_3)_2 \cdot NH_4NO_3 \cdot H_2O$	15.5			19				20
Calcium cyanamide	$Ca(CN)_2$	20-22			37				63
Sodium nitrate	$NaNO_3$	16							29
Urea	$CO(NH_2)_2$	45-46							
Methylene urea		38-41							
Urea ammonium nitrate	$NH_4NO_3 \cdot CO(NH_2)_2$	28-32							
Phosphorus Materials									
Single superphosphate	$Ca(H_2PO_4)_2$		18-20		18-21		12	—[3]	—
Triple superphosphate	$Ca(H_2PO_4)_2$		45-26		12-14		1	—	—
Phosphoric acid	H_3PO_4		52-54					110	
Superphosphoric acid	—[2]		76-83					160	
Potassium Materials									
Potassium chloride	KCl			60-62				—	—
Potassium nitrate	KNO_3	13		44-46					26
Potassium sulfate	K_2SO_4			50-53			18	—	—
Potassium thiosulfate	$K_2S_2O_3$			25			17	26	
Potassium magnesium sulfate	$K_2SO_4 \cdot MgSO_4$			33	0.1	11	18	—	—

[1]Equivalent per 100 lb of nutrient (N, P_2O_5 and K_2O). For example, 216 lb of urea (100 lb N) has the equivalent acidity of 80 lb of $CaCO_3$.

[2]H_3PO_4, $H_2P_2O_7$, $H_5P_3O_{10}$, $H_6P_4O_{13}$, and other higher-phosphate forms.

[3]The symbol '-' refers to a neutral reaction, neither acidic nor basic.

MICRONUTRIENTS

Properties of the various micronutrient fertilizers vary considerably. A micronutrient material may be completely water soluble or only slightly soluble. Nutrient availability increases with solubility. Inorganic sources may be relatively pure compounds or mixtures of compounds containing one or more micronutrients. Micronutrients can also be reacted with carbon containing compounds such as synthetic chelates or natural organic complexes of metal ions. Some micronutrients are available as insoluble metal oxides and low solubility metal oxysulfates. Insoluble and low solubility micronutrients provide a slow source of micronutrients or, when ground to a very fine particle size, are available for direct plant nutrient uptake. Classifications of micronutrient sources include inorganic salts, synthetic chelates, and natural organic (carbon compound) complexes.

Inorganic Salts

Sulfates of copper, iron, manganese, and zinc, plus borates and molybdates, are the most common sources of inorganic micronutrients. The most commonly used boron material is sodium tetraborate, which is relatively water-soluble. Soluble polyborate, boric acid, and sodium octaborate are often used as foliar sprays.

Water-soluble ammonium and sodium molybdate are the primary sources of molybdenum. Molybdic oxide, a slightly soluble compound, is occasionally used.

See Table 7-6 for composition and properties of inorganic micronutrient materials.

Synthetic Chelates

A chelating agent is a compound that combines with a polyvalent metal cation to form a protective, open ring structure around the metal cation. Chelation delays precipitation of the metal ions in the soil by neutralizing the positive charge so metal cations are not retained on the cation exchange complex, nor do they react with hydroxides in the soil solution. Chelation also increases mobility of metal micronutrients in soil and soilless media. Commercially available synthetic chelating agents and the concentration of micronutrients are shown in Table 7-7 (page 192).

TABLE 7-6
Inorganic Sources of Micronutrients

Material	Formula	Element (%)	Water Solubility (g/100g H_2O)	°F
Sources of Boron				
Boric acid	H_3BO_3	10.0	6.3	80
Granular borax	$Na_2B_4O_2 \cdot 10H_2O$	11.3	2.5	33
Sodium tetraborate, anhydrous	$Na_2B_4O_7$	21.5	1.3	32
Solubor®	$Na_2B_8O_{13} \cdot 4H_2O$	20.5	22.0	86
Ammonium pentaborate	$NH_4B_5O_8 \cdot 4H_2O$	19.9	7.0	64
Sources of Copper				
Copper sulfate	$CuSO_4 \cdot 5H_2O$	25.0	24.0	32
Cuprous oxide	Cu_2O	88.8	Insoluble	
Cupric oxide	CuO	79.8	Insoluble	
Cuprous chloride	Cu_2Cl_2	64.2	1.5	77
Cupric chloride	$CuCl_2$	47.2	71.0	32
Sources of Iron				
Ferrous sulfate	$FeSO_4 \cdot 7H_2O$	20.1	33.0	32
Ferric sulfate	$Fe_2(SO_4)_3 \cdot 7H_2O$	19.9	440.0	68
Iron oxalate	$Fe_2(C_2O_4)_3$	30.0	Very Insoluble	
Ferrous ammonium sulfate	$Fe\,(NH_4)_2(SO_4)_2 \cdot 6H_2O$	14.2	18.0	
Ferric chloride	$FeCl_3$	34.4	74.0	32
Sources of Manganese				
Manganous sulfate	$MnSO_4 \cdot 4H_2O$	24.6	105.0	32
Manganous carbonate	$MnCO_3$	47.8	0.0065	77
Manganous oxide	MnO_4	72.0	Insoluble	
Manganous chloride	$MnCl_2$	43.7	63.0	32

(continued)

TABLE 7-6 (Continued)

Material	Formula	Element (%)	Water Solubility (g/100g H_2O)	°F
Manganous oxide	MnO	77.4	Insoluble	
Sources of Molybdenum				
Sodium molbydate	$Na_2MoO_4 \cdot H_2O$	39.7	56.0	32
Ammonium molybdate	$(NH_4)_6Mo_7O_{24} \cdot 4H_2O$	54.3	44.0	77
Molybdic oxide	MoO_3	66.0	0.11	64
Sources of Zinc				
Zinc sulfate	$ZnSO_4 \cdot H_2O$	36.4	50.0	60
Zinc oxide	ZnO	80.3	Insoluble	
Zinc carbonate	$ZnCO_3$	52.1	0.001	60
Zinc chloride	$ZnCl_2$	48.0	432.0	77
Zinc oxysulfate	$ZnO \cdot ZnSO_4$	52.0	1.3	60
Zinc ammonium sulfate	$ZnSO_4 \cdot NH_4SO_4 \cdot 6H_2O$	16.3	9.6	32
Zinc nitrate	$Zn(NO_3)_2 \cdot 6H_2O$	22.0	324.0	68

TABLE 7-7
Synthetic Chelates

Chelating Agent	Micronutrients Content, % Element[1]			
	Cu	Fe	Mn	Zn
EDTA	7–13	5–14	5–12	6–14
HEEDTA	4–9	5–9	5–9	9
NTA		8		13
DTPA		10		
EDDHA		6		

[1]Where a range is shown, the low number indicates the analysis of common liquid forms and the high number indicates the analysis of common dry forms.

Formation constants and chelate-metal stability over pH ranges are important criteria to consider when evaluating different chelating agents. Generally, the stability of metal chelates is greater near neutral than at low or high pH values. This is an important consideration for the formulator incorporating metal chelates into macronutrient fertilizers. For example, if Zn-EDTA is mixed with phosphoric acid prior to ammoniation, the complex will break down, but if Zn-EDTA is mixed with an ammoniating solution, the complex will remain stable.

Natural Organic Complexes

Many naturally occurring compounds contain chemically reactive groups similar to synthetic chelating agents. Examples include lignosulfonates, polyflavonoids, polysaccharides, amino acids, and humic acids. Compounds used commercially to complex micronutrients are often prepared from by-products of other industries. Metal complexes with these organic compounds have lower stability than synthetic chelates. Microorganisms in soil more readily break down organic complexes. Most are suitable for foliar sprays and for mixing with fluid fertilizers.

MAXIMIZING NUTRIENT EFFICIENCY

Considerable effort has been directed toward developing fertilizers to overcome specific problems or to fulfill the nutrient needs of specific plants. In the past, these products were considered "specialty fertilizers," but today slow release, controlled release, and stabilized fertilizers are widely used in and beyond the horticultural industry. For simplicity, the term "slow release" will be used in this text to reference all three forms unless specified. Slow release fertilizers are intended to maximize nutrient uptake efficiency while minimizing the potential for negative environmental effects, such as leaching and runoff. Slow release fertilizers reduce the risk of surface and groundwater pollution without sacrificing performance. The wide range of nutrient release characteristics available from slow release fertilizers provide growers with the flexibility to better match plant nutrient requirements. The grower can select an appropriate slow release fertilizer for a specific plant type and growing cycle.

In 1985, the Association of American Plant Food Control Officials (AAPFCO) defined *slow* or *controlled release fertilizer* as such:

> "A fertilizer containing a plant nutrient in a form which delays its availability for plant uptake and use after application, or which extends its availability to the plant significantly longer than a reference 'rapidly available nutrient fertilizer' such as ammonium nitrate or urea, ammonium phosphate, or potassium chloride. Such delay of initial availability or extended time of continued availability may occur by a variety of mechanisms. These include controlled water solubility of the material (by semi-permeable coatings, occlusion, or by inherent water insolubility of polymers, natural nitrogenous organics, protein materials, other chemical forms), by slow hydrolysis of water soluble low molecular weight compounds, or by other unknown means."

A distinction can be made between slow and controlled release fertilizers in that a controlled release fertilizer (CRF) is intended to provide plant nutrition over a very specific time period under a certain set of environmental conditions, whereas a slow release fertilizer has a less defined period of release. The marketplace generally considers reacted nitrogen products to fall within the category of slow release fertilizer and coated materials to fall within the realm of controlled release fertilizer.

To maximize fertilizer efficiency, the release rate should match the nutrient needs of the plant over time. Matching plant needs with nutrient inputs will improve plant quality and appearance while providing safety and labor savings through reduced applications of fertilizers.

To date, there are several approaches to achieving slow release characteristics:

- The development of compounds that have limited water solubility
- The addition of a water-soluble physical barrier

- The addition of a non-soluble physical barrier that allows nutrients to release through osmosis or diffusion
- The use of chemical additives or inhibitors to slow the transformation of nitrogen resulting in an extended period of availability, known as *stabilized nitrogen*

Nitrogen, the most widely applied fertilizer nutrient, is the most susceptible to loss by leaching or denitrification. Nitrogen fertilizers, therefore, make up the majority of slow release fertilizers. Some common slow release fertilizers are listed in Table 7-8. Even though these fertilizers cost more initially than conventional fertilizers, they are widely used for improved nutrient efficiency, overall cost savings including labor, and better plant response with less top growth. Potential SRF markets include golf courses, lawn and landscape care companies, consumer lawn and garden centers, nurseries, greenhouses, flower growers, and producers of many other ornamental plants.

TABLE 7-8

Analyses of Some Slow Release Fertilizers

Material	Method of Release	Nitrogen Content (%)
Resin/polymer-coated N-P-K	Osmotic barrier	Varies by substrate and release longevity
Resin/polymer-coated N	Osmotic barrier	37 – 44
Sulfur-coated urea	Coating with sulfur	32 – 38
Polymer coated/ sulfur-coated urea	Osmotic/sulfur barrier	37 – 43
Isobutylidene diurea	Solubility	31
Methylene urea	Solubility/microbial	38 – 41
Urea-formaldehyde	Microbial	35 – 38
Melamine	Solubility/microbial	66
Stabilized nitrogen	Enzyme & microbial suppression	46
Resin/polymer-coated KNO_3	Osmotic barrier	42 – 44 (K_2O)

The development of slow release fertilizers is a fairly recent occurrence, with the basic timeline as follows:

1955	Urea-formaldehyde commercially introduced
1960–1970	Methylene ureas developed
1960s	Polymer-coated CRF developed and introduced
1960–1970	Sulfur-coated ureas developed and commercially introduced
1980s	Methylene diurea/dimethylene triurea composition developed, and additional polymer-coated fertilizers introduced
1990s	Polymer/sulfur-coated fertilizers introduced

TYPES OF SLOW RELEASE FERTILIZERS

Nitrogen Reaction Fertilizers

Nitrogen reaction fertilizers are produced by a chemical reaction of water-soluble nitrogen sources, such as urea or ammonia, to produce more complex molecular structures. These fertilizers have a limited water solubility that controls the nitrogen availability to plants.

Urea-Formaldehyde Fertilizers

Urea-formaldehyde (UF) fertilizers are formed by the reaction of urea with formaldehyde. This reaction produces nitrogen-containing polymers of varying molecular weights or chain lengths. The longer the polymer chain, the less soluble it is, which extends the time of release.

Granular compositions of UF fertilizers are divided into three classes, each based on the degree of water solubility as affected by polymer length:

- Ureaform is the least water-soluble
- Methylene urea has intermediate water-solubility
- Methylene diurea/dimethylene triurea (MDU/DMTU) is the most water-soluble

Each class of UF products contains certain amounts of unreacted urea. MDU/DMTU compositions, the lowest molecular-weight compounds, contain the greatest amount of unreacted urea nitrogen, and ureaform contains the least amount.

Ureaform, as defined by the American Association of Plant Food Control Officials (AAPFCO), is sparingly soluble. *Ureaform* contains a minimum of 35 percent total nitrogen, with at least 60 percent of it as Cold Water Insoluble Nitrogen (CWIN). It must also have an *Activity Index (AI),* the percentage of CWIN that is soluble in hot (100° C) water, of not less than 40 percent. Ureaform is composed largely of longer-chain urea formaldehyde polymers, primarily tetramethylene pentaurea (TMPU) and longer. The unreacted urea nitrogen content is usually less than 15 percent of the total nitrogen. Ureaform is used in spikes and stakes for houseplants and planting tabs for ornamentals.

Methylene ureas are a class of sparingly soluble fertilizers that contain predominantly intermediate chain-length polymers, primarily trimethylene tetraurea (TMTU) and tetramethylene pentaurea (TMPU). The total nitrogen content of these polymers is 39 to 40 percent, with between 25 and 40 percent of the nitrogen present as CWIN. The unreacted urea content generally is in the range of 15 to 20 percent of the total nitrogen. This class of granular ureaform products is not specifically defined under the AAPFCO "Official Terms and Definitions."

MDU/DMTU compositions, the newest class of methylene urea fertilizers, were developed in the 1980s. These granular compositions are predominantly shorter-chain polymers with at least 60 percent of their polymeric nitrogen in the form of the cold-water-soluble polymers methylene diurea (MDU) and dimethylene triurea (DMTU). These polymers have a total nitrogen content of above 40 percent, with generally less than 25 percent of their total nitrogen as CWIN. Because of the shift to shorter-length polymers, these fertilizers contain higher percentages of unreacted urea than ureaforms and methylene ureas.

Ureaformaldehyde solutions are clear water solutions. They contain very low molecular weight, water-soluble urea-formaldehyde reaction products, plus unreacted urea. Various concentrations of the ureaformaldehyde solution are produced. They contain a maximum of 55 percent unreacted urea, with the remainder as one or more of methylolureas, methylolurea ethers, or triazone.

Urea-Other Aldehyde Fertilizers (31-0-0)

Isobutylidene diurea (IBDU) is a combination of urea and isobutyraldehyde. This reaction forms a single oligomer, which is a white crystalline solid. The nitrogen from IBDU becomes available to plants through hydrolysis. IBDU will hydrolyze to urea and isobutyraldehyde, with the rate increasing under acid conditions and high temperatures. Because it is not dependent on microbes, IBDU nitrogen is available at lower temperatures than other nitrogen reaction fertilizers. The dependence on moisture for nitrogen release is a distinguishing characteristic of IBDU.

Crotonylidene diurea (CDU) is produced from urea being acid catalyzed with crotonaldehyde or acetaldehyde. CDU is a white powder. The nitrogen is available to plants through a combination of hydrolysis and microbial decomposition. CDU will degrade more rapidly in warmer, acidic conditions. It is used primarily in Japan and Europe, where it is produced. Its use is limited in the United States.

Coated Fertilizers

Coated fertilizer is a term characterizing those products in which a water-soluble fertilizer granule is covered with a coating. The coating limits or controls the rate of water penetration to the soluble fertilizer core and, in some products, controls the release rate of solubilized fertilizer from within the granule to the soil.

The amount of material exhibiting controlled release properties may vary from 40 to 98 percent. The variability comes from different coating materials and coating methods used. The four categories of coated fertilizers are those that:

- Use primarily sulfur as the coating material
- Employ a polymeric material
- Are hybrid products with a multilayer coating of sulfur and polymer
- Are fertilizers with a fused magnesium phosphate coating

The market growth of coated fertilizers is the result of more favorable economics, increased flexibility in nutrient release patterns as compared to nitrogen reaction products, and the opportunity to control the release of other nutrients in addition to nitrogen.

Sulfur-Coated Urea

Sulfur-coated urea (SCU) is produced by coating a urea granule with molten sulfur. Sulfur was chosen as a coating material for its low cost and for its value as a secondary nutrient. Normally, a soft wax is used as a secondary coating over the sulfur to fill imperfections and to provide handling integrity to the brittle sulfur coat. The nitrogen content of SCUs (30 to 40 percent) varies with the amount of coating applied, along with the sulfur content (up to 17 percent).

The method of release of SCU is by water penetration through micropores and imperfections in the coating. A rapid release of dissolved urea from the core of the particle takes place through these cracks or imperfections. Microbes degrade the wax and the sulfur coating. Excess impact and abrasion during shipping, blending, and application can result in increased cracking of the protective sulfur coat, thereby reducing the longevity of release.

The release rate of SCUs is affected by the coating thickness and the coating quality. Particles with thicker sulfur coating have a longer release period. The risk is that particles with too much sulfur coating will exhibit lock-off and will release over very long periods (i.e., two years or more). The manufacturing process, however, leads to a blend of lightly to heavily coated urea granules typically releasing nitrogen during a 6 to 12-week period.

SCUs are generally used in fertilizer blends to supply nitrogen. Longevity of a finished blend is a function of the amount of SCU in the blend (15 to 100 percent) and the application rate used. The largest use of SCUs is in turf markets, with limited use in nursery and specialty agricultural markets.

Polymer-Coated Sulfur-Coated Urea

Polymer-coated sulfur-coated ureas (PCSCU) are products that utilize granular urea with a primary coating of molten sulfur and a secondary polymer coat. These fertilizers were developed to deliver better nutrient release performance than SCUs at a lower cost than polymer coated urea. Like SCUs, sulfur is used as a primary coating because of its lower cost. Thin layers of a polymer overcoat are then used to further improve handling durability. Unlike the SCUs with a soft wax sealant, the PCSCU products use polymers to provide a more durable outer coating over the brittle sulfur coating. If left unprotected, the brittle sulfur coating may be damaged through handling. Nitrogen content may vary from 35 to 43 percent, depending on the coating thickness used.

The method of release for PCSCUs is through a combination of hydrolysis, microbiological activity, and capillary action. Water must first penetrate through the continuous polymeric membrane layer. The rate of penetration through the membrane depends on the composition and thickness of the polymeric film. Water then penetrates through defects in the sulfur coat by capillary action and solubilizes the fertilizer core. As the fertilizer solubilizes, the pressure inside the granule increases until the granule bursts and releases the nitrogen into the soil. The durability of the polymer coating increases the performance of these products through better handling characteristics.

Similar to the SCUs, PCSCUs are generally used in fertilizer blends to supply nitrogen. When used in turf markets, PCSCUs are generally blended with uncoated constituents. In nursery markets, they may be blended with additional polymer-coated fertilizers.

Polymer-Coated Fertilizers

Polymer-coated fertilizer (PCF) is produced by using heat to coat a fertilizer substrate with either a thermoset or thermoplastic resin. Polymer coatings are used to coat single-nutrient compounds, such as urea, or ammonium-based N-P-K substrates, and potassium nitrate. Secondary nutrients and micronutrients may also be incorporated into the substrate or be mixed into the coating to extend their utility. Depending on the coating used, coating weights vary from 1.5 to 20 percent of the fertilizer. Generally, PCF coatings are harder and will withstand abrasion better than the SCUs and PCSCUs.

The method of release for PCFs is through diffusion. Water must diffuse through the continuous polymeric membrane, with the rate of diffusion dependent on the composition and thickness of the polymeric membrane. Water then solubilizes the fertilizer core and diffuses back across the membrane through osmotic pressure differentials. This diffusion-controlled mechanism permits the greatest uniformity of nutrient release, with release periods of 3 to

36 months. The release period, referred to as longevity, is dependent upon the amount and type of coating applied, as well as the substrate used. Blends of varying product longevities give uniform release over longer periods. Polymer-coated fertilizers are dependent on temperature for release, where the release rate increases as the temperature increases.

In the past, because of their high cost of production, PCFs had been sold primarily for high-value applications, such as in container nurseries, landscaping, and strawberry production. However, improved technology introduced in the early 1990s has lowered the cost of production of PCFs to the point where they are now also used extensively in the turf market.

Factors Affecting Release Rates

Factors affecting release rates from the various controlled release fertilizers discussed in this chapter are summarized in Table 7-9.

NUTRIENT CONVERSION FACTORS

Nutrients in fertilizers are reported in elemental form, with the exception of phosphorus and potassium. Several foreign countries and most scientific literature report all nutrients on an elemental basis. Often it is necessary to determine the percentage of an element in a fertilizer. Table 7-10 (page 203) provides the information needed to make the conversion from compounds to elements and from elements to compounds for many fertilizers. To use this table, one would need the label of the fertilizer in consideration. Here is one example of how to use this table. Suppose one has an ammonium phosphate fertilizer (16-20-0), which contains 20 percent P_2O_5 and one needs to determine the percent of elemental phosphate (P) in that substance. Multiply 20 percent $P_2O_5 \times 0.4364 = 8.73$ percent P. The 0.4364 conversion factor was obtained from column "A to B" in Table 7-10.

Likewise, to find the percent MgO in a 9 percent Mg fertilizer, one would multiply $9 \times 1.6579 = 14.9$ percent MgO.

TABLE 7-9

Parameters That Affect Mechanisms of Nitrogen Release Among Different Slow Release Fertilizers

						Coating Characteristics		
Fertilizer	**Temperature**	**Bacteria**	**Moisture**	**pH**	**Particle Size**	**Thickness**	**Chemical Composition**	**Durability**
Natural Organics	***/****	****	***	*	**	N/A	N/A	N/A
Longer-Chained UF	***/****	***/****	*	*	—	N/A	N/A	N/A
Shorter Chained UF	**/***	**	**	**	*	N/A	N/A	N/A
Isobutylidene Diurea	*/**	*	***	*/**	****	N/A	N/A	N/A
Stabilized Nitrogen	**	*	*	*	N/A	N/A	**	N/A
Polymer-Coated Sulfur-Coated Ureas	**	*	**	—	**	**	**	***
Polymer-Coated Fertilizers	***	—	*	—	***	***	**/***	***

Parameters that affect the mechanism of release are rated from none to high. Some Parameters are not applicable (N/A) to the appropriate category. Ratings are from: - none, * slight, ** moderate, ***high, **** very high.

TABLE 7-10

Nutrient Conversion Factors

To get from A to B, multiply A amount by the factor in column "A to B" and vice versa.

A	B	Multiply A to B	Multiply B to A
Amonium (NH_3)	Nitrogen (N)	0.8224	1.2159
Nitrate (NO_3)	Nitrogen (N)	0.2259	4.4266
Protein (crude)	Nitrogen (N)	0.1600	6.2500
Protein (wheat)	Nitrogen (N)	0.1754	5.7000
Ammonium nitrate (NH_4NO_3)	Nitrogen (N)	0.3500	2.8572
Ammonium sulfate [$(NH_4)_2SO_4$]	Nitrogen (N)	0.2120	4.7168
Calcium ammonium nitrate solution (CAN–17) [$Ca(NO_3)_2 \cdot NH_4NO_3$]	Nitrogen (N)	0.1707	5.8572
Ammonium calcium nitrate decahydrate ("Calcium nitrate" dry) [$5Ca(NO_3)_2 \cdot NH_4NO_3 \cdot 10H_2O$]	Nitrogen (N)	0.1550	6.4516
Potassium nitrate (KNO_3)	Nitrogen (N)	0.1386	7.2176
Sodium nitrate ($NaNO_3$)	Nitrogen (N)	0.1648	6.0679
Monoammonium phosphate ($NH_4H_2PO_4$)	Nitrogen (N)	0.1218	8.2118
Diammonium phosphate [$(NH_4)_2HPO_4$)]	Nitrogen (N)	0.2121	4.7138
Urea	Nitrogen (N)	0.4665	2.1437
Phosphoric acid (P_2O_5)[1]	Phosphorus (P)	0.4364	2.2914
Phosphate (PO_4)	Phosphorus (P)	0.3261	3.0662
Monoammonium phosphate ($NH_4H_2PO_4$)	Phosphoric acid (P_2O_5)[1]	0.6170	1.6207
Diammonium phosphate [$(NH_4)_2HPO_4$)]	Phosphoric acid (P_2O_5)[1]	0.5374	1.8607

(continued)

TABLE 7-10 (Continued)

A	B	Multiply A to B	Multiply B to A
Monocalcium phosphate [$Ca(H_2PO_4)_2$]	Phosphoric acid (P_2O_5)[1]	0.6068	1.6479
Dicalcium phosphate ($CaHPO_4 \cdot 2H_2O$)	Phosphoric acid (P_2O_5)[1]	0.4124	2.4247
Tricalcium phosphate [$Ca_3(PO_4)_2$]	Phosphoric acid (P_2O_5)[1]	0.4581	2.1829
Potash (K_2O)	Potassium (K)	0.8301	1.2046
Muriate of potash (KCl)	Potash (K_2O)	0.6317	1.5828
Sulfate of potash (K_2SO_4)	Potash (K_2O)	0.5405	1.8499
Potassium of nitrate (KNO_3)	Potash (K_2O)	0.4658	2.1466
Potassium carbonate (K_2CO_3)	Potash (K_2O)	0.6816	1.4672
Gypsum ($CaSO_4 \cdot 2H_2O$)	Calcium sulfate ($CaSO_4$)	0.7907	1.2647
Gypsum ($CaSO_4 \cdot 2H_2O$)	Calcium (Ca)	0.2326	4.3000
Gypsum ($CaSO_4 \cdot 2H_2O$)	Calcium oxide (CaO)	0.3257	3.0702
Calcium oxide (CaO)	Calcium (Ca)	0.7147	1.3992
Calcium carbonate ($CaCO_3$)	Calcium (Ca)	0.4004	2.4973
Calcium carbonate ($CaCO_3$)	Calcium oxide (CaO)	0.5604	1.7848
Calcium carbonate ($CaCO_3$)	Calcium hydroxide [$Ca(OH)_2$]	0.7403	1.3508
Calcium hydroxide [$Ca(OH)_2$]	Calcium (Ca)	0.5409	1.8487
Magnesium oxide (MgO)	Magnesium (M)	0.6032	1.6579
Magnesium sulfate ($MgSO_4$)	Magnesium (M)	0.2020	4.9501
Epsom salt ($MgSO_4 \cdot 7H_2O$)	Magnesium (M)	0.0987	10.1350
Sulfate (SO_4)	Sulfur (S)	0.3333	3.0000
Ammonium sulfate [$(NH_4)_2SO_4$]	Sulfur (S)	0.2426	4.1211
Gypsum ($CaSO_4 \cdot 2H_2O$)	Sulfur (S)	0.1860	5.3750
Magnesium sulfate ($MgSO_4$)	Sulfur (S)	0.3190	3.1350
Potassium sulfate (K_2SO_4)	Sulfur (S)	0.1837	5.4438

A	B	Multiply	
		A to B	B to A
Sulfuric acid (H_2SO_4)	Sulfur (S)	0.3269	3.0587
Borax ($Na_2B_4O_7 \cdot 10H_2O$)	Boron (B)	0.1134	8.8129
Boron trioxide (B_2O_3)	Boron (B)	0.3107	3.2181
Sodium tetraborate pentahydrate ($Na_2B_4O_7 \cdot 5H_2O$)	Boron (B)	0.1485	6.7315
Sodium tetraborate anhydrous ($Na_2B_4O_7$)	Boron (B)	0.2150	4.6502
Cobalt nitrate [$Co(NO_3)_2 \cdot 6H_2O$)	Cobalt (Co)	0.2025	4.9383
Cobalt sulfate ($CoSO_4 \cdot 7H_2O$)	Cobalt (Co)	0.2097	4.7690
Cobalt sulfate ($CoSO_4$)	Cobalt (Co)	0.3802	2.6299
Copper sulfate ($CuSO_4$)	Copper (Cu)	0.3981	2.5119
Copper sulfate ($CuSO_4 \cdot 5H_2O$)	Copper (Cu)	0.2545	3.9293
Ferric sulfate [$Fe_2(SO_4)_3$]	Iron (Fe)	0.2793	3.5804
Ferrous sulfate ($FeSO_4 \cdot H_2O$)	Iron (Fe)	0.3150	3.1746
Ferrous sulfate ($FeSO_4 \cdot 7H_2O$)	Iron (Fe)	0.2009	4.9776
Manganese sulfate ($MnSO_4$)	Manganese (Mn)	0.3638	2.7486
Manganese sulfate ($MnSO_4 \cdot 4H_2O$)	Manganese (Mn)	0.2463	4.0602
Sodium molybdate ($Na_2MoO_4 \cdot 2H_2O$)	Molybdenum (Mo)	0.3965	2.5218
Sodium nitrate ($NaNO_3$)	Sodium (Na)	0.2705	3.6970
Sodium chloride (NaCl)	Sodium (Na)	0.3934	2.5417
Zinc oxide (ZnO)	Zinc (Zn)	0.8034	1.2447
Zinc sulfate ($ZnSO_4$)	Zinc (Zn)	0.4050	2.4693
Zinc sulfate ($ZnSO_4 \cdot H_2O$)	Zinc (Zn)	0.3643	2.7450

[1]Also called phosphoric acid anhydride, phosphorus pentoxide, available phosphoric acid.

SUPPLEMENTARY READING

Association of American Plant Food Control Officials. www.aapfco.org

Centre International des Engrais Chimique (CIEC). 1968. *New Fertilizer Materials*. Noyes Development Corp., Park Ridge, NJ.

Engelstad, O. P., T. J. Army, J. J. Hanway, and V. J. Kilmer, ed. 1985. *Fertilizer Technology and Use*. Soil Science Society of America, Madison, WI.

Follett, R., et al. 1981. *Fertilizers and Soil Amendments.* Prentice-Hall, Englewood, NJ.

Goertz, H. M. 1993. *Controlled Release Technology*. p. 251–274. Kirk-Othmer. *Encyclopedia of Chemical Technology*. 4th ed. Vol. 7. John Wiley, New York, NY.

Khasawneh, F. E., E. C. Sample, and E. J. Kamprath, ed. 1980. *The Role of Phosphorus in Agriculture.* American Society of Agronomy. Madison, WI.

Landels, S. P. 1991. *Proceedings of the Controlled Release Fertilizer Workshop*. p. 87–101. National Fertilizer and Environmental Research Center, TVA.

Landels, S. P., M. M. Smart, J. Bakker, and J. Shimosato. 1989. Controlled release fertilizers and nitrification inhibitors. *Chemical Economics Handbook*. SRI International. Menlo Park, CA.

Mortvedt, J. J., F. R. Cox, L. M. Shuman, and R. M. Welch, ed. 1991. *Micronutrients in Agriculture*, 2nd ed. Soil Science Society of America, Madison, WI.

Munson, R. D., ed. 1985. *Potassium in Agriculture*. American Society of Agronomy, Madison, WI.

Shoji, S. and A. Gandizer. 1992. *Controlled Release Fertilizers with Polyolefin Resin Coating*. Konno Printing, Sendai, Japan.

Stevenson F. J., ed. 1982. *Nitrogen in Agricultural Soils*. American Society of Agronomy, Madison, WI.

Tisdale, S. L., W. L. Nelson, J. D. Beaton, and J. L. Havlin. 1993. *Soil Fertility and Fertilizers,* 5th ed. The Macmillan Company, New York, NY.

Windholz, M., ed. 1983. *The Merck Index*, 10th ed. Merck and Co. Inc., Rahway, NJ.

CHAPTER 8

Organic Sources of Nutrients

There are many reasons a grower selects organic or unprocessed nutrients for plant production. Some growers want to use locally available nutrient resources for economic reasons. Others may be producing plants that will be certified for the organic market. Whatever the purpose, this chapter will focus on the information needed for those growers who are incorporating organic sources of nutrients into their plant production systems.

Plant nutritional needs can be met with a variety of organic materials, but the behavior of these materials is often more difficult to predict than with manufactured, synthetic fertilizers. Most organic nutrient sources depend on biological or chemical processes for breakdown and nutrient release. These processes are governed by complex interactions of environmental, soil, and management practices. Many growers familiar with single element manufactured fertilizers may apply organic materials to supply one or two plant nutrients while, in fact, organic materials may contain a wide variety of essential nutrients. Many organic sources do not contain nutrients in the ratios required by plants, so caution should be used to avoid undesirable accumulation of unutilized nutrients in the soil.

The selection of specific nutrient sources should be based on cost, availability, plant nutritional needs, and the requirements of certifying agencies. Remember that the fundamental principles of plant nutrition, soil biology, and plant growth remain the same regardless of the nutrient source. There is value in testing new organic products as they become available. However, one must maintain proper soil nutrient levels for achieving desired levels of plant growth.

The use of approved nutrient sources is governed by a variety of regional, national, and international oversight organizations. Each organization maintains its own standards and allows different materials to be used in their organic production systems. As a result, growers must consider the organic input requirements of each program for which certifications will be requested.

The National Organic Program in the United States and the Canadian General Standards Board classifies products as either allowed, restricted, or prohibited for use in organic production. "Allowed" products are permitted for organic production when applied as directed on the label. "Restricted" materials can only be applied for certain uses and under specific conditions. "Prohibited" products may never be used for organic production. The properties and values of these materials as sources of plant nutrients vary considerably.

NITROGEN SOURCES FOR ORGANIC PRODUCTION

Nitrogen (N) is the plant nutrient that is most limiting to efficient and profitable plant growth. An inadequate supply of available nitrogen frequently results in plants that have slow growth, poor color, undesirable appearance, and inefficient water use. Nitrogen-stressed plants often have greater disease susceptibility compared with properly nourished plants. However, excessive nitrogen can be detrimental to plant growth and quality, and may cause undesirable environmental impacts. For these reasons, more research has been conducted on managing this plant nutrient than any other. This chapter does not address all the important aspects related to nitrogen management, but covers the major sources of nitrogen for organic plant production and their behavior in soil. More detailed information can be found in Chapter 2.

Earth's atmosphere contains 78 percent nitrogen gas, but most organisms cannot directly use this resource due to the stability of the compound. Breaking the strong chemical bond in nitrogen gas (N_2) requires either the input of energy in commercial fertilizer production, or biological nitrogen fixation by specialized nitrogenase enzymes. Applications of manufactured nitrogen fertilizers are not allowed in organic production, so high energy nitrogen fixation is not discussed in this chapter.

There are many biological and chemical processes that impact the plant uptake of nitrogen. Low nitrogen uptake efficiency can be caused by an imbalance of essential plant nutrients. Management of nitrogen is made difficult due to uncertainties related to weather events following fertilization. When nitrogen uptake is low, it is important to consider the fate of the unused nitrogen.

Almost all non-legume plants obtain nitrogen from the soil in the form of inorganic ammonium (NH_4^+) or nitrate (NO_3^-). Some organic nitrogen containing compounds can be acquired by roots in small amounts, but these are not a major source of plant nutrition. Some plants, especially grasses, grow better with an ammonium nitrogen source. However, microbial nitrification processes rapidly oxidize ammoniacal nitrogen to nitrate. Most plants grow best with predominantly nitrate-nitrogen nutrition. In most warm, well-aerated soils, the nitrate concentration may be at least 10 times greater than the ammonium concentration.

Unlike other plant nutrients, there is no universal or widely used soil test to predict the amount of required supplemental nitrogen. Instead, nitrogen supplementation is typically based on desired plant appearance, past management, and measurement of residual soil or media nitrate. Most processed fertilizers contain soluble nitrogen sources so nitrogen availability to plants is quite predictable. On the other hand, most organic nitrogen sources require mineralization before they can be utilized by plants. Environmental factors, such as soil temperature, pH, and moisture all impact the rate of nitrogen availability from organic sources.

A major factor for using organic nitrogen sources involves knowing both the amount of nitrogen applied and the rate of nitrogen release from the organic material in the year after application. Nitrogen availability coefficients, also called *plant-available-nitrogen (PAN),* are used to estimate the fraction of total nitrogen that will be available for plant uptake during the first growing season. Nitrogen availability coefficients can vary widely based on the nature of the material, management practices, and environmental factors. Examples of PAN coefficients are shown in Table 8-1.

Mineralization of Organic Matter

When the plant's nitrogen supply comes exclusively from organic sources, such as green manure and composts, a thorough understanding of mineralization is essential to avoid a deficiency or surplus of available nitrogen. Mineralization rates are not consistent through the year. Mineralization rates are dependent on environmental factors, such as temperature and soil moisture, the properties of the organic material—such as C:N ratio and lignin content—and the placement of the material. Many references, found in the Supplementary Reading section at the end of this chapter, discuss mineralization in more detail.

Ideally, nitrogen release from mineralization should match plant nitrogen demand. Figure 8-1a illustrates a situation

TABLE 8-1

Examples of Plant Available Nitrogen Coefficients

		Surface Applied	
	Soil Incorporated	**Broadcast**	**Irrigated**
Manure Type	**Fraction of N Available During the First Year**		
Poultry litter	0.6	0.5	—
Layer manure	0.6	0.4	—
Scraped swine manure	0.6	0.4	—
Scraped dairy manure	0.6	0.4	—
Swine lagoon effluent	0.8	0.5	0.5
Dairy lagoon effluent	0.8	0.5	0.5
Compost (C:N 15-20:1)	0.05	0.03	
Compost (C:N > 25:1)	0	0	

Baldwin, K.R., & J.T. Greenfield. 2009. *Composting on Organic Farms.*

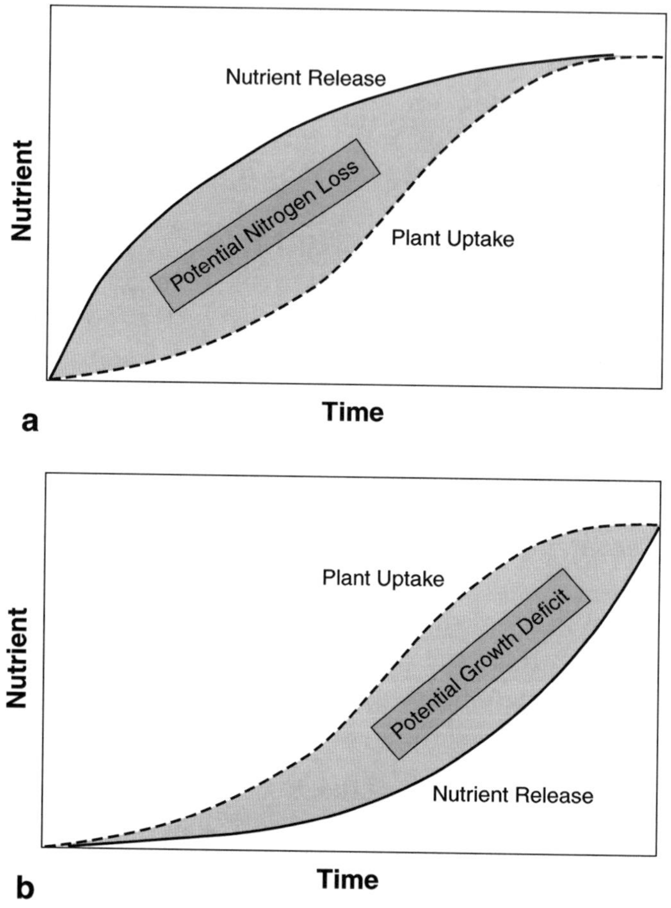

FIGURE 8-1 a) Scenario where nitrogen mineralization is faster than plant uptake and; b) mineralization lags behind plant demand.

where nitrogen mineralization occurs faster than plant uptake, leading to excess soil nitrate-nitrogen that has the potential to be lost by leaching, denitrification, or other means. Figure 8-1b describes a situation where nitrogen mineralization is slower than plant nitrogen demand, which leads to potential plant nitrogen deficiency.

Composts

Composts contain relatively low concentrations of nitrogen, phosphorus, and potassium. Composts typically decompose slowly and behave as a slow-release source of nitrogen. It takes many months or years for composts to break down since the rapidly decomposable compounds have been previously degraded during the composting process. Composts can be made from on-site materials, but they are also widely available from municipal and manufactured sources.

Composts vary in quality and tend to have low immediate nutritional value, but provide valuable sources of stable organic matter. Since municipal compost may contain plastic, trash, and industrial waste, some organic certification programs do not allow their use. Commercial compost is widely available from a variety of organic sources.

Manure

The chemical, physical, and biological properties of fresh *manure* vary tremendously due to specific animal feeding and manure management practices. Manure nitrogen is present in both organic and inorganic forms. Nitrogen is unstable in fresh manure because ammonia gas can be readily lost through volatilization. Application of fresh manure or slurry on the soil surface can result in volatilization losses as high as 50 percent of the total nitrogen. The combination of wet organic matter and nitrate in some manure can also result in significant denitrification losses. The organic nitrogen-containing compounds in manure become available for plant uptake following mineralization by soil microorganisms, while the inorganic nitrogen fraction is immediately available. Figure 8-2 shows the wide range in nitrogen mineralization rates from 107 individual dairy manure samples after eight weeks of incubation. On average, 13 percent of the organic nitrogen was mineralized, but 13 samples had net immobilization. Net nitrogen mineralization from the remaining 94 samples ranges from 0 to 55 percent.

Manures and composts can be challenging to uniformly apply to a planted area due to their bulky nature and inherent variability. Application of raw manure may bring up concerns, such as pathogens, hormones, and medications. The use of raw manure is restricted for some organic uses and growers should check with the certifying agency before using.

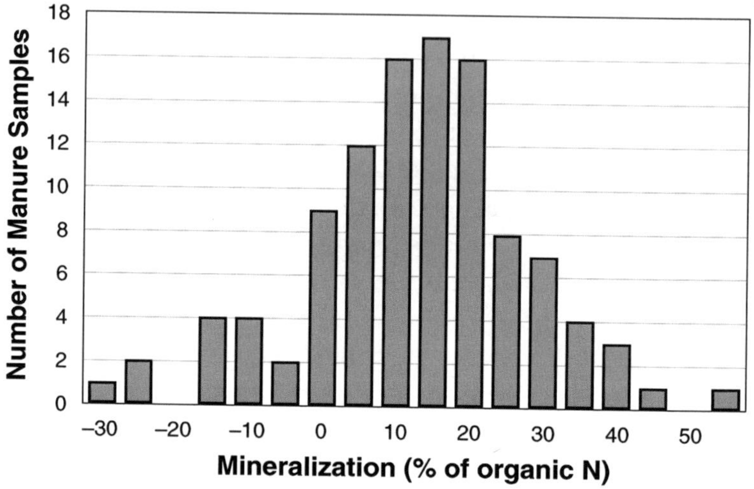

FIGURE 8-2. The distribution of organic nitrogen mineralization rates from dairy manure. *Source:* Van Kessel and Reeves. 2002. *Biology an Fertility of Soils*. 36:118–123.

Organic Nitrogen Fertilizers

Plant Products

Alfalfa meal (4 percent N), cottonseed meal (6 percent N), corn gluten (9 percent N), and soybean meal (7 percent N) are all examples of plant products that are sometimes used as nitrogen sources for organic production. These products are also used as protein-rich animal feeds. They require microbial mineralization before the nitrogen is available for plant uptake. Mineralization of these nitrogen-rich materials is generally rapid.

Animal By-Products

Dried, powdered *blood meal* (12 to 14 percent N) is derived mainly as a waste by-product from the slaughter of cattle. Dried blood is completely soluble, rapidly mineralizes to plant-available forms, and is suitable for distribution through irrigation systems after it has been solubilized.

Seabird guano (8 to 12 percent N) is derived from natural deposits of excrement and remains of birds living along extremely arid seacoasts. Guano was historically a very important nitrogen source before industrial processes for making fertilizer were developed. Many of the major guano deposits are now exhausted. Guano is also harvested from caves where large bat populations roost. It can be applied directly to soil or dissolved in water to make a liquid suspension fertilizer.

Feather meal (14 to 16 percent N) consists of finely ground poultry feathers and is a major by-product of the poultry industry. Feather meal may contain between 70 to 90 percent protein. Feather protein is mostly present as non-soluble keratin stabilized by highly resistant disulfide bonds. When treated with pressurized steam and animal-derived enzymes, the feather-based protein becomes a good source of available nitrogen for plant nutrition. Much of the feather nitrogen is not initially soluble, but it mineralizes relatively quickly under conditions favorable for plant growth. Pelletizing the feather meal makes handling and application more convenient. Unprocessed feathers usually have a delayed nitrogen release, but can also be an excellent nitrogen source if the difficulty in uniformly applying low-density feathers to the soil can be overcome.

Both dry *fish meal* (10 to 14 percent N) and liquid *fish emulsion* (2 to 5 percent N) are processed waste by-products derived from fish wastes. Non-edible fish are cooked and pressed to separate the solid and liquid fractions. The solids are used as fish meal for fertilizer and animal feed. The fish oil is removed from the liquid fraction and the remaining solution is thickened into fish emulsion. Additional processing is often performed to prevent premature decomposition. The odor from fish meal products may be unpleasant in a closed environment such as a greenhouse. Mineralization of fish-based products is generally rapid.

High nitrogen animal by-products have relatively rapid nitrogen mineralization. When these organic fertilizers are incorporated into soils at temperature between 50°F and 80°F, more than half of the organic nitrogen may mineralize within two weeks of application (Figure 8-3).

Organically certified sodium nitrate fertilizer (16 percent N) is mined from naturally occurring deposits in arid regions of Chile and Peru where nitrate salts have accumulated over time. Sodium nitrate is generally granulated and readily soluble when added to soil. Sodium nitrate is more likely to be used in organic agriculture

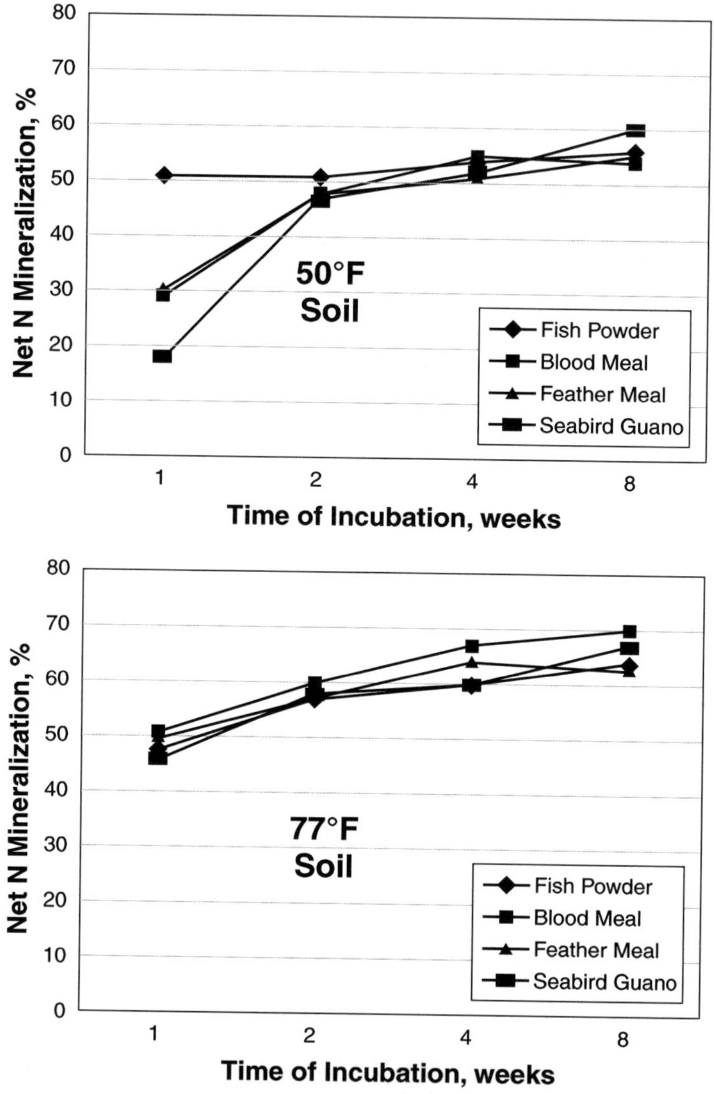

FIGURE 8-3. Nitrogen mineralization of four common organic nitrogen fertilizers at two soil temperatures. Mineralization of nitrogen expressed as percent of added organic nitrogen. *Source:* Hartz, T.K., and R. Johnstone. 2006. *Hort Technology*. 16:39-42.

and is not often used in turf and ornamental plant production. The use of sodium nitrate is restricted by some certification programs.

Seaweed-based products are typically derived from kelp species, mainly from the genus *Ascophyllum* harvested from the northern Atlantic Ocean. Dried kelp contains approximately 1 percent nitrogen and 2 percent potash (K_2O), as well as small amounts of other plant nutrients. Due to their low nutritive content, kelp products are generally used for reasons of growth promotion rather than plant nutrition.

Choosing the "best" source of nitrogen for organic plant production is difficult since plant demand, fertilizer nutrient ratios, plant-available nitrogen, mineralization rates, local access, ease of application, acceptance of source by certifying organizations, and cost all need to be considered. Computer-based tools are available to help with these choices. For example, Oregon State University has an "Organic Fertilizer Calculator" program that allows comparison of various materials to best meet the fertility needs of plants.

Each organic nitrogen source has a unique characteristic that requires special management to maximize plant health and economic production, while minimizing undesirable environmental losses. Commercial organic sources tend to be more expensive than inorganic nitrogen sources, but many local or on-site nitrogen sources may also be available. Some locally available nitrogen sources may contain low concentrations of nitrogen, requiring transportation and handling of large volumes of material. As the understanding of soil nitrogen and organic matter improves, better nitrogen management will benefit growers and the environment.

ORGANIC PHOSPHORUS FERTILIZERS

Nutrient management in organic production systems focuses on maintaining productivity with inputs of readily available or minimally processed materials. Nutrient inputs for organic production are typically focused on carbon-based nutrient sources such as plant residue, compost, manure, and non-processed mineral sources such as rock phosphate, lime, and gypsum.

The nutrient reservoir in the soil shrinks when plants are removed from a managed area and not reincorporated, such as bagging and disposing grass clippings rather than mulching. This nutrient export creates a phosphorus deficit, necessitating regular phosphorus additions to replace the removed phosphorus. Because phosphorus is an essential nutrient for plant growth, all sustainable systems should, at a minimum, seek to replace the phosphorus removed in harvested plants in order to avoid declines in yield and quality. Although organic plant production seeks to minimize imported inputs, it is essential that producers replace phosphorus removed when plant residues are not returned to the soil.

Soil organic matter can be an important source of phosphorous for plants. Some, but not all studies have shown that soil organic matter increases when nutrients are managed organically. These differences depend on many soil management practices such as return of plant residues and climatic factors. Soil organic matter serves as a reservoir of plant nutrients and improves the soil's physical and bio-chemical conditions in the root zone.

Soil organic matter contains a variety of organic phosphorus compounds, such as inositol phosphate, nucleic acid, and phospholipids. These compounds must be first converted to inorganic phosphate by soil enzymes before being used for plant growth. Phosphatase enzymes are produced by soil microorganisms, mycorrhizal fungi, or excreted by plant roots. Some organic phosphorus compounds are stable for many years in the soil, while others are converted to inorganic phosphorus within a few days or weeks.

Mycorrhizal Fungi

Soil fungal organisms that form a symbiotic association with plant root cells are known as *mycorrhizal fungi*. These fungi invade plant root cells and derive essential carbohydrate nutrition from the plant. In return, mycorrhizal fungi provide benefits to plants by extending large mats of fungal hyphae that greatly expands a plant's root system. Phosphorus is taken up by plants through diffusion. Since phosphorus is not mobile in the soil solution, an extensive root system enhances phosphorus uptake. Almost all plants grown in soil form this relationship with mycorrhizal fungi, which is present in the root zone of most mineral soils. Figure 8-4 shows a mycorrhizal association with roots.

FIGURE 8-4. Plant roots with mycorrhizal fungi.

Many organic growers encourage the association of mycorrhizal fungi with plant roots through the use of inoculated seed during turf establishment, as well as surface application of mycorrhizal preparations after aerification. Frequent tillage may disrupt the soil fungal network and may reduce its effectiveness for providing nutrients to the plant.

The value of mycorrhizal fungi for supplying phosphorus to plants is most apparent in low phosphorus soils. In most cases, plants growing in soils with medium to high concentrations of phosphorus have less mycorrhizal association than plants in low phosphorus conditions. Therefore, the value of mycorrhizal fungi is greatest in soils without an adequate supply of phosphorus. Mycorrhizal fungi do not provide any additional phosphorus to the soil, but can allow better utilization of the existing soil resource. Commercial sources of mycorrhizal fungi are available and may be used in specialized conditions.

Rock Phosphate

Rock phosphate is a general term used to describe a variety of globally distributed phosphorus-rich minerals. Of the two main types, sedimentary or igneous, sedimentary rock deposits are the source of over 80 percent of the total world production of phosphate rock. Depending on its geologic origin, rock phosphate has widely varying mineralogy, texture, and chemical properties. Some rock phosphate is found in hard rock deposits, while other rock phosphorus is found as soft, soil-like material. The great variation in properties and the accompanying elements present in the rock, such as carbonate and fluoride, have a large effect on the value of rock phosphate as a source of plant nutrients. The information on phosphorus availability from a specific rock source is generally not available to the consumer.

The general reaction of rock phosphorus dissolution added to soils to become a plant available form is:

$$Ca_5(PO_4)_3F + 6H^+ \leftrightarrow 5Ca^{2+} + 3H_2PO_4^- + F^-$$

Note that this is a reversible reaction, so the driving forces for dissolution of rock phosphate are high acidity (H^+) and low calcium (Ca^{2+}).

It is difficult to make universally applicable recommendations for rock phosphate application because so many factors affect its dissolution and plant availability. However, there are some key factors to consider.

Soil pH is important in the dissolution of rock phosphate. Rock phosphate is much more soluble in acidic soils (soil pH <5.5). In neutral pH to alkaline soils, rock phosphate typically provides little benefit for plant nutrition.

Phosphorus particle size influences the dissolution of rock phosphate by controlling the surface area available for reaction. However, some rock phosphate is so insoluble that fine grinding will not significantly increase phosphorus availability. Conversely, it may not be necessary to finely grind highly reactive rock phosphate used for direct application to the soil. Many rock phosphate sources are commonly ground to finer than 100 mesh (0.15 mm) to improve reactivity, but such finely ground material may be difficult to handle and to spread uniformly.

Low soil calcium concentrations and high soil cation exchange capacity favor rock phosphate dissolution since calcium is one of the reaction products resulting from dissolution. Soil conditions that limit calcium availability, such as soil acidity, high leaching, or the presence of organic compounds that complex exchangeable calcium, tend to favor rock phosphate dissolution and the release of phosphorus to the plant.

Other cultural practices that may improve phosphorus availability from rock phosphate are broadcast applications that maximize soil dissolution reactions, and using management methods that promote root colonization by mycorrhizal fungi. Application of rock phosphate should be made between several weeks to months prior to the anticipated need for plant nutrients. Although lime applications are important for reducing harmful effects associated with soil acidity, lime additions tend to reduce the value of rock phosphate as a nutrient source by increasing soil pH and soluble calcium.

Manure and Composts

These materials are good sources of phosphorus for plants. Even though these materials are considered organic products, over 75 percent of the total phosphorus they contain is present as inorganic compounds. A general rule suggests phosphorus in manure and compost is 70 percent available for soils with low soil-test phosphorus, but 100 percent available for soils testing adequate or high for phosphorus.

The ratio of nutrients in composts and manures does not closely match that required by plants. When manure and compost are used as a primary nitrogen source for plants, phosphorus is typically over-applied by three to five times the required rate. Long-term use of manures and compost as the primary nitrogen source leads to an accumulation of phosphorus in the soil that can become an environmental concern for surface water quality.

Bone Meal

Bone meal, prepared by grinding animal bones, is one of the earliest phosphorus sources used in agriculture. Most commercially available bone meal is steamed to remove any raw animal tissue. The primary phosphorus mineral in bone material is calcium-deficient hydroxyapatite, which is more soluble than rock phosphate, but much less soluble than conventional phosphorus fertilizers. Calcium-deficient hydroxyapatite present in bone meal dissolves as such:

$$Ca_{9.5}(HPO_4)_{0.5}(PO_4)_{5.5}(OH)_{1.5} + 13H^+ \leftrightarrow 9.5Ca^{2+} + 6H_2PO_4^- + 1.5H_2O$$

Similar to rock phosphate, bone meal is most effective in acidic soils and when the particle size is small. When used properly, it can be an effective phosphorus source. Reacting animal bones with sulfuric acid to enhance the solubility of phosphorus produced one of the first commercial phosphorus fertilizers.

Concerns have been raised regarding bovine spongiform encephalopathy (BSE) in cattle and the residual effect of bone meal as a fertilizer. However, there are no restrictions on the use of bone meal. Most commercial bone meal products have been heat treated, reducing the potential for disease transmission.

Bird and Bat Guano

Guano is most commonly used as a source of nitrogen for plants, but some guano materials are also relatively rich in phosphorus. Guano is mined from aged deposits of bird or bat excrement in low rainfall environments. The drying and aging process changes the chemistry of the phosphorus compared with fresh manure. Struvite (magnesium ammonium phosphate) can be found in substantial quantities in guano. However, struvite dissolves slowly in soil. The limited supply and high cost of guano generally restricts its use to small-scale applications.

There are several options available for meeting the phosphorus requirement for organic production. Growers are encouraged to first consider locally available materials to meet this need. Many of the allowed materials are fairly low in nutrient content. Therefore, transportation costs are a concern since relatively large quantities of amendment are needed to meet plant demand. Regular soil and tissue testing should be conducted by all growers to avoid depletion of soil nutrients and to prevent inadvertent nutrient accumulation, regardless of production philosophy and management techniques.

ORGANIC SOURCES OF POTASSIUM

Potassium (K) is an essential nutrient for plant growth, but it often receives less attention than nitrogen and phosphorus in many organic production systems. Many regions of the United States and all of the Canadian provinces remove more K during harvest than is returned to the soil in fertilizer and manure. This results in a depletion of nutrients from the soil and increasing occurrences of deficiency in many locations.

Potassium is the soil cation required in the largest amount by plants, regardless of nutrient management philosophy. Large amounts of potassium are required to maintain plant health and vigor. Plant-available potassium is usually measured in the topsoil, but some deep-rooted plant species can take up considerable amounts of potassium from the subsoil. Maintenance of adequate potassium is essential for both organic and conventional plant production. Refer to Chapter 2 for more information on the roles of potassium in plant growth and development.

Supplemental potassium is sometimes called "potash," a term that comes from an early production technique where potassium was leached from wood ashes and concentrated by evaporating the leachate in large iron pots. Clearly this practice is no longer practical and is not environmentally sustainable. This potash collection method depended on the tree roots to acquire soil potassium, which was then recovered after the wood was harvested and burned. Most potassium fertilizer, whether used in organic or conventional agriculture, comes from ancient marine salts, deposited as inland seas evaporated. This natural geological process is still visible in places such as the Great Salt Lake and the Dead Sea.

In general, regulations for organic potassium sources specify that they must not be processed, purified, or altered from their original form. However, there is disagreement between different certifying bodies over what specific materials can be used.

Regular applications of soluble potassium, regardless of the source, will increase the concentration of potassium in the soil solution and the proportion of potassium on the cation exchange sites. The commonly used soluble potassium sources, including manure, compost, and green manure, contain this nutrient in the simple cationic (K^+) form. Most soluble inorganic fertilizers and organic manure are virtually interchangeable as sources of potassium for plant nutrition. Any differences in plant performance are usually due to the accompanying anions, such as chloride (Cl^-) or sulfate (SO_4^{2-}), or the organic matter that may accompany the added potassium.

There is no general evidence that potassium sulfate (K_2SO_4) is more effective than potassium chloride (KCl) as a source of plant-available potassium. Both sulfate and chloride provide essential nutrients that are required for plant health. Chloride has a well-documented role in improving plant health and the prevention of a variety of plant diseases. Chloride-derived salinity was the same as sulfate-based salinity on its effect on common soil microbes in laboratory research. The addition of potassium decreased the harmful effects of salinity on soil microbial activity.

Approved and Restricted Organic Potassium Sources

Langbeinite (potassium magnesium sulfate; $K_2SO_4 \cdot MgSO_4$) is allowed as a nutrient source if it is used in the raw, crushed form without any further refinement or purification. Several sources of langbeinite are available for use in organic plant production. Langbeinite typically contains 18 percent K_2O, 11 percent Mg, and 22 percent S in forms readily available for plant uptake. The major source of langbeinite in North America is from underground deposits in New Mexico.

Greensand is the name commonly applied to a sandy rock or sediment containing a high percentage of the green mineral glauconite. Because of its potassium content (up to 5 percent K_2O), greensand has been marketed for over 100 years as a natural fertilizer and soil conditioner. The very slow potassium release rate of greensand is touted to minimize the possibility of plant damage by fertilizer "burn", while the mineral's moisture retention may aid soil conditioning. However, the potassium release rate is too slow to provide any significant nutritional benefit to plants at realistic application rates. Soluble potassium is generally less than 0.1 percent of the total potassium present. Deposits of greensand are found in several states, including Arkansas and Texas, but the only active greensand mine in North America is located in New Jersey.

Since manure and compost are extremely variable, based on their raw materials and their handling, they also contain highly variable potassium concentrations. Composted organic matter is generally allowed as a nutrient source. Raw manures have restrictions on the timing of their use, but the details depend on the certifying agency. The potassium in these organic materials is largely available for plant uptake, similar to approved inorganic sources. Repeated applications of large amounts of manure can result in potassium accumulation in the soil, which may lead to excessive uptake of potassium by the plant beyond plant demand. A chemical analysis of the manure or compost composition is necessary in order to use these resources for maximum benefit.

When potassium sulfate is derived from natural sources, it is allowed for organic plant production. Much of the current production of organically approved potassium sulfate in North America comes from the Great Salt Lake in Utah. Organic potassium sulfate may not undergo further processing or purification after mining or evaporation, other than crushing and sieving. This product is not allowed in some European countries without special permission from the certifying agency. It generally contains approximately 50 percent K_2O and 17 percent S.

Mined rocks, biotite, mica, feldspars, granite, and greensand are allowed without restriction. Variability exists in the potassium release rate from these mineral sources. Some are unsuitable potassium sources for plant nutrition due to their limited solubility and their heavy and bulky nature. In general, a smaller particle size translates to a greater surface area, reactivity, and weathering rate. As with all organic nutrient sources, check with the organic certifying agency before making any purchases.

Since sea water contains an average of 0.4 ppm potassium, seaweed may accumulate up to several percent potassium. When harvested, seaweed biomass can be used directly as a potassium source or the soluble potassium may be extracted. These potassium sources are readily soluble and typically contain less than 2 percent K_2O. While seaweed-derived products are excellent potassium sources, their low potassium content and high transportation costs can make it problematic for large area use, especially when the source is far from the point of application.

Sylvinite is a naturally occurring, mined potassium chloride mineral. Sylvinite use is restricted in the USDA standards unless it is from a mined source and undergoes no further processing. It must be applied in a manner that minimizes chloride accumulation in the soil. Generally, sylvinite should only be used after consultation with the certifying agency. Unprocessed sylvinite contains approximately 17 percent K_2O.

Wood ash from hardwood trees served as one of the earliest sources of potassium. This highly variable material is composed of the elements initially present in the wood that were not volatilized when burned. Wood ash is an alkaline material with a pH ranging from 9.0 to 13.0 and has a liming effect between 8 and 90 percent of the total neutralizing value of commercial limestone. In terms of commercial fertilizer, average wood ash would have an analysis of approximately 0 percent N, 1 percent P_2O_5, and 4 percent K_2O. The use of ash derived from manures, biosolids, and coal is prohibited for organic production. Check with the certifying organization prior to applying ash to soil.

Growers using organic production practices, like all growers, have the need for an adequate supply of soil potassium to sustain healthy plants. There are many excellent sources of potassium that are available for replacing the nutrients removed from the soil in harvested plants. Failure to maintain adequate potassium in the root zone will result in poor water use efficiency, greater pest problems, decreased harvest quality, and reduced plant quality. Regular soil testing for potassium is the key for establishing the requirement for fertilization. If a need for supplemental potassium exists, organic producers should first consider locally available potassium resources and supplement with mineral sources. The expense of transporting and applying low nutrient content amendments must also be considered.

SUPPLEMENTARY READING

Andrews, N., and J. Foster. 2007. *Organic Fertilizer Calculator*. Oregon State University, Corvallis, OR. EM. 8936-E smallfarms.oregonstate.edu/organic-fertilizer-calculator.

Baldwin, K.R., and J.T. Greenfield. 2009. *Organic Production Composting on Organic Farms*. www.cefs.ncsu.edu/resources/organicproductionguide/compostingfinaljan2009.pdf. Center for Environmental Farming Systems.

Gaskell, M., and R. Smith. 2007. Nitrogen sources for organic vegetable crops. *Horticulture Technology* 17: 431–441.

Hartz, T.K., and R. Johnstone. 2006. Nitrogen availability from high-nitrogen-containing organic fertilizers. *Horticulture Technology* 16: 39–42.

Li, X., F. Li, B. Singh, and Z. Rengel. 2006. Decomposition of maize straw in saline soil. *Biology and Fertility of Soils* 42: 366–370.

Mikkelsen, R.L. 2007. Managing potassium for organic crop production. *Horticulture Technology* 17: 455–460.

Mikkelsen, R.L. et al. 2008. Understanding and managing N-P-K for Organic Crop Production. *Better Crops with Plant Food* Vols. 1, 2, and 4. International Plant Nutrition Institute. www.ipni.net/organic/references.

Okur, N., M. Cengal, and S. Gocmez. 2002. Influence of salinity on microbial respiration and enzyme activity of soils. *Acta Horticulture* 573: 189–194.

Sullivan, D.M. 2008. *Estimating Plant Available Nitrogen from Manure*. Oregon State University, Corvallis, OR. EM-8954E. extension.oregonstate.edu/catalog/pdf/em/em8954-e.pdf.

Van Kessel, J.S., and J.B. Reeves III. 2002. Nitrogen mineralization potential of dairy manures and its relationship to composition. *Biology and Fertility of Soils* 36: 118–123.

Various Authors. Department of Soil Science, College of Agricultural and Life Sciences. North Carolina State University. www.soil.ncsu.edu/about/publications.php#AnimalWaste.

CHAPTER 9

Methods of Applying Fertilizers

Fertilizers are used to provide nutrients that are not present in the soil or the planting medium in amounts necessary to meet the needs of growing plants. When choosing the method of fertilizer application, growers should consider the following:

- Root characteristics of the plant species
- Plant nutrient requirements at different stages of growth and/or desired growth characteristics
- Physical and chemical characteristics of the soil or planting media
- Physical and chemical characteristics of the fertilizer material to be applied
- Moisture availability
- Irrigation method

Production of horticultural plants may require multiple applications of nutrients. Thus, several fertilizer application methods may be employed in the same area depending on the needs of the plant. For example, established turfgrass may receive a spring and a fall application of a complete fertilizer using a broadcast applicator. During the growing season, to maintain turf quality and color, additional nitrogen may be applied through an irrigation system or a broadcast spreader.

Fertilizers added to soils or media undergo transformations that may change their chemical or positional availability. The methods of application are related to the utilization of plant nutrients and the changes the nutrients undergo when placed in the growing media or soils. The appropriate application method should be as economical, accurate, and efficient as possible.

This chapter discusses different methods of fertilizer application, the equipment used, and the benefits and limitations of each application technique.

SURFACE APPLICATION

Broadcast application of fertilizers consists of uniformly distributing dry or liquid fertilizer over the desired area.

A *drop spreader* is a simple, manual applicator for dry fertilizer consisting of an inverted triangular-shaped hopper mounted between two wheels (Figure 9-1, left). Fertilizer is distributed through adjustable openings in the bottom of the hopper. When the spreader is properly calibrated, very accurate application rates are obtainable.

A *broadcast spreader* is a dry fertilizer spreader consisting of a bin mounted over a horizontal spinning disc (Figure 9-1, right, and Figure 9-2). The fertilizer drops through adjustable openings in the bottom of the bin and the spinning disc broadcasts it over a wide area. Proper calibration of the spreader permits a reasonably accurate and fast application.

Large dry fertilizer spreaders are available, each consisting of a bin mounted on a two or four-wheeled trailer frame and pulled by a tractor, truck, or all-terrain vehicle (ATV). The spreaders are powered by a power take-off (PTO) or ground driven power that works off the wheels. The spreader can also be mounted to a tractor with a three-point hitch. The fertilizer is usually spread in a 20 to 40 foot swath by a horizontal spinning disc. Prevention of skips or excessive overlaps of fertilizer is essential.

FIGURE 9-1. Drop spreader (left) and broadcast spreader (right). Both are manual push-type spreaders used for small areas.

Surface applications of liquid fertilizers are made through irrigation systems and sprayers. The basic components of a liquid fertilizer broadcast applicator includes a tank, a pressure gauge and regulator, a pump, pipes, hoses, fittings, a boom, and nozzles. The applicator can be mounted on a truck, a flotation vehicle, an all-terrain vehicle, a trailer, or directly on a tractor.

FIGURE 9-2. A power take-off (PTO) [shown] or ground driven rotary spreader is often used for rapid coverage on golf courses or other large turf areas.

During the growing season, containers and pots may have fertilizer applied to the top portion of the media, which is known as *topdressing*. Topdressing in turfgrass refers to applications of amendments, such as sand, compost, etc.

SUBSURFACE APPLICATION

The term *soil injection* refers to the application of fertilizers below the surface of the soil. Metal shanks are used to create channels in the soil, with drop pipes or flexible tubes following. Typically, drop pipes are used when injecting liquid fertilizers, while flexible polyethylene tubing is used when injecting dry fertilizers. Fertilizers that can be broadcast onto the surface of the soil can also be injected, but there is very limited use of this method of fertilizing in the turf and ornamental industry.

There are several potential advantages to using soil injection rather than surface broadcast application when applying fertilizers. Certain nitrogen fertilizers may be subject to gaseous losses when left on the soil surface. If soils are subject to wind or water erosion, injection helps to prevent nutrient losses by placing fertilizers below the zone subject to erosion. Immobile nutrients, such as phosphorus, potassium, and some forms of micronutrients, can be placed directly into the root zone. This is beneficial because the fertilizer is in contact with less substrate, resulting in less fixation and enhanced nutrient availability. Additionally, fertilizer may be placed deep enough in moist soil zones where feeder root activity is greatest. Soil injection is particularly well suited to field grown plants whose roots explore a limited amount of soil. Placement directly into the root zone reduces fertilizer availability to weeds and minimizes application to the portions of the soil with few active roots.

Soil injection application also presents several disadvantages. Power requirements are greater than for surface broadcast application and fewer acres can be treated in a given amount of time. Additionally, if fertilizer is applied on very wet or heavy soils, injection tools may disrupt the physical integrity of the soil.

Subsurface bands of fertilizer may be placed to the side of and/or below the seed or established plant. The applicator can be set to place the fertilizer bands at any depth or position relative to the seed or plant. The use of tablets or fertilizer spikes, like those used in landscaping and container stock, would be defined as subsurface application.

Caution: Fertilizer placed too close to the seed or established plant may damage roots or inhibit seed germination. Placement with the seed is not generally recommended, unless a high quality controlled release fertilizer is used.

Incorporation refers to mixing fertilizers into containers or potting mixes prior to planting or seeding. The media mixes are blended in commercial blenders, in concrete-type blenders, or on pads or benches. The fertilizer is mixed uniformly throughout the medium.

FERTIGATION

Fertigation—the application of fertilizer in irrigation water—saves time, labor, and fuel costs. Applications may be pre-plant or post-emergence, using either liquid or dry soluble fertilizer materials. Application of fertilizer through irrigation limits the distribution uniformity to the consistency of the irrigation system, but increases the probability that the fertilizer will be placed into the root zone. Every effort should be made to prevent the movement of fertilizer-laden water off-site. Fertigation can be used in high and low volume sprinkler systems, trickle and drip systems, and through furrow and flood irrigation.

Not all dry and liquid fertilizers are suitable for application in closed systems. Due to relatively large orifice sizes and high flow rates, high-pressure sprinkler systems require less refinement in fertilizer formulation and application technique. In these systems, most liquid fertilizers may be used, with the exception of strong acids, aqua ammonia, and anhydrous ammonia. The use of acid fertilizers and amendments requires systems constructed of corrosion-resistant materials. Aqua ammonia and anhydrous ammonia are not generally recommended for use in closed systems because precipitates may form if the irrigation water contains a high concentration of calcium. Plant foliage injury may result from sprinkler application of dilute ammonia solutions; thus, this method of fertilizer application should be avoided if working with decorative foliage and ornamentals.

Filters are advised when applying fertilizers through closed irrigation systems, particularly low-volume microirrigation systems. Low operating pressures, non-turbulent flow, and small orifice size make these systems more susceptible to plugging than high-volume sprinkler systems. To reduce the opportunity for plugging, fertilizers should be introduced into a system at a point well ahead of the filters. Automated filter systems back flush when particulates reduce flow. Back flush water is typically discharged into a pond or ditch. It is important to either install an automated switch or manually stop fertilizer injection when filters back flush to prevent loss of fertilizer into the back flush water. Small in-line filters may be installed in the tubing that connects the field storage tank with the injection apparatus. An in-line filter removes solid material that may be present in the fertilizer tank or transfer truck. Only clean base stock solutions and uncoated dry fertilizers should be used in formulations for low-volume microirrigation systems.

Various injectors are available for introducing fertilizer materials into closed irrigation systems (Figure 9-3). Metering pumps inject the fertilizer into the main irrigation line under pressure at a constant rate. Proportioning pumps possess flow-monitoring devices that adjust fertilizer injection rates based on water flow rates so that a constant fertilizer concentration is maintained. Both metering and proportioning pumps may be powered by gas or electric motors, or by water-driven pumps that derive their energy from the main irrigation pump. The pressure developed by these pumps must be greater than the pressure within the closed irrigation system in order for injection to occur. Conversely, *venturi-type injectors* introduce fertilizer by creating a vacuum that draws the fertilizer into the system. Venturi-type injectors are very simple devices that involve no moving parts. A short length of rigid tubing is attached to the main irrigation line at two points separated by several feet. The tubing is constricted for a short distance, which increases the velocity of water as it passes through this constriction. A connector hose from the fertilizer tank is attached at the point of constriction. A pressure drop accompanies this increase in velocity. A 20 percent

FIGURE 9-3. A liquid fertilizer storage and injection system.

pressure differential is sufficient to initiate a negative pressure (vacuum), which draws liquid fertilizer from the connector hose.

Special tanks are available for dissolving dry fertilizers for application through closed systems. Solution injection machines are capable of injecting dissolved gypsum directly into irrigation systems. The injection system consists of a fiberglass tank containing agitation and recirculation devices. Finely ground gypsum is added to the tank and mixed with irrigation water diverted through the tank. The water, enriched with dissolved gypsum, is then returned to the main irrigation line. Injection systems of this type are a simple way to introduce dry fertilizers into closed irrigation systems. Filters are provided to remove insoluble components.

For any system, it is important that the injector be of adequate size to provide the quantity of fertilizer that must be applied based on unit of flow, area, or time. Backflow-prevention devices must be used to keep fertilizer materials from being accidentally introduced into the water source.

As a special precaution in a microirrigation system, the entire system should be thoroughly flushed with clean water after injection of fertilizers. A simple practice is to inject fertilizer in the middle portion of an irrigation application, allowing for system pressurization at the beginning and flushing at the end. This effectively prevents accidental mixing of incompatible materials within the irrigation system.

When water drips onto the soil surface, the wetted zone represents about 1 to 2 percent of the total wetted soil surface. Consequently, all fertilizer chemical reactions are concentrated by about 50 to 100 times in terms of their concentration with drip irrigation.

Before injecting an unknown fertilizer into irrigation water, it is strongly recommended to perform a jar test to check for compatibility. The jar test procedure is conducted as such:

- Fill a glass gallon container with irrigation water
- Add two to three drops of the fertilizer to be injected into the water and mix
- After about 1 hour of standing, examine the container for any scum at the surface, any sediment at the bottom, and rotate it in sunlight to observe any occurring opalescence
- If any of these three conditions occur, do not inject the fertilizers and contact your fertilizer dealer to identify what has caused the problem

If two or more fertilizers are injected at the same time, the grower is advised to use separate dedicated injectors for each individual chemical. This reduces the chance of chemical reactions from occurring and plugging the lines. Injection ports should be located about three feet apart along the irrigation line to ensure complete mixing prior to injecting the next fertilizer.

Most well waters and some surface waters contain appreciable concentrations of bicarbonate (HCO_3^-) ions. Very high pH (> 8.3) water may also contain carbonate (CO_3^{2-}) ions. *Alkalinity* is the term used when referring to the combined concentrations of bicarbonate and carbonate ions in water. These ions can cause precipitation reactions when fertilizers are injected into irrigation water. Phosphate fertilizers are especially susceptible to precipitation in high bicarbonate water. Bicarbonate concentrations greater than 2 meq/L (120 ppm) should be amended by injecting sulfuric acid, urea sulfuric acid, or sulfur dioxide gas (SO_2) to reduce the water pH to 6.5, thereby removing about 50 percent of the existing bicarbonates. If urea sulfuric acid is used, the amount of urea-nitrogen must be deducted from the overall nitrogen budget. Many commercial laboratories can perform a titration analysis using the desired acid to determine the correct amount of acid to inject to achieve a target water pH.

Some greenhouse managers continuously inject phosphoric acid to lower water pH to 4.0 to provide phosphate nutrition and lower the possibility of fertilizer precipitation. However, low pH (< 4.0) reduces the solubility and availability of molybdate (MoO_4^{2-}). Molybdenum is essential for nitrogen transformations in plants, so an apparent nitrogen deficiency may result when irrigating with highly acidic water.

Growers may benefit from injecting ammonium polyphosphate fertilizer at a rate of 2 ppm phosphorus in the irrigation water. This represents the addition of only 40 milliliters of liquid ammonium polyphosphate (10-34-0) fertilizer per 1,000 gallons of water. Do not use liquid orthophosphate fertilizers because this will plug the irrigation lines with the formation of insoluble calcium phosphate. The ammonium polyphosphate is not sufficient to supply all of the needs of the crop. However, it will ensure about 1 to 2 ppm phosphorus in solution around the roots, which is much greater than normally present. Consequently, phosphorus will no longer be limited, the plants will usually grow larger, and the total nutrient need of these larger plants will require adjusting the total fertilizer program to compensate for the improved growth.

Never use anhydrous or aqua ammonia with a drip system because free ammonia gas can enter and kill or retard the growth of plants. The extremely high pH of these fertilizers will also precipitate calcium carbonate, which will plug drip emitters.

Never inject phosphate fertilizer with calcium or magnesium fertilizers together because this will result in a very insoluble residue forming inside the lines. It is difficult to dissolve this residue. Immediate application of 2 to 5 gallons per acre (6 to 15 oz per 1,000 sq ft) of urea sulfuric acid has been shown to help unplug drip emitters following co-injection of calcium and phosphorous.

Fertigation with drip irrigation works ideally with the various urea, ammonium, and nitrate fertilizers. The grower can fine-tune the amount of nitrogen applied on a regular basis, preventing excessive leaching of nitrate and promoting better plant growth. Plant nitrogen needs change over the growing season and can be matched with varying injection rates.

Micronutrient metal ions (iron, manganese, copper, zinc, and nickel) can be injected into irrigation lines. It is best to apply these micronutrient metal ions to the soil through fertigation in the form of chelated metals. Do not inject any form of copper into an aluminum irrigation line—because copper acts to dissolve the inside of the aluminum pipe and leaks will quickly develop.

A very serious problem often develops with permanent drip fertigation, especially on sandy low cation exchange capacity (CEC) soils. The use of urea and ammonium fertilizers causes the accumulation of strong acidity in the top 6 inches below the drip zone. The soil pH often will become 3.0 to 4.5 and create a toxic level of aluminum ions (Al^{3+}). The roots will turn brown to black and will be killed below the drip zone. Applying calcium carbonate lime to the soil surface will be ineffective because lime must come into direct contact with the acid for it to neutralize the acidity.

Use caution when using "instant lime" materials such as potassium carbonate and potassium hydroxide. These materials do instantly lime the acidic zone. However, the amount required to create neutralization overloads the soil CEC sites with potassium ions (K^+) acting similar to sodium ions (Na^+) causing serious soil dispersion and reducing water penetration. Once all of the acidity has been neutralized, these "instant lime" materials cause the soil pH to zoom upward to pH 12.0 or higher. This results in dissolving the soil humus and causes a permanent destruction of the soil structure. It is best to apply these materials in small volumes repeatedly over time, checking soil pH in between applications. Always follow label recommendations.

Precipitation of calcium phosphates is influenced by the source of phosphorus used. The use of furnace or purified-grade phosphoric acid is generally recommended for microirrigation systems. As an alternative to the use of furnace or purified-grade orthophosphoric acid, injection of sulfuric acid or urea sulfuric acid prior to injection of polyphosphate solutions has been used to prevent calcium phosphate precipitates. Achieving a pH of greater than 4.5, but less than 5.5, during injection of phosphorus fertilizers greatly reduces the risk of forming calcium phosphate precipitates. Lowering the pH to 3.0 for a short period of time after application of phosphorus will dissolve precipitates present within the system. Prolonged exposure to acidic water may damage aluminum, brass, uncoated carbon steel, and other acid sensitive components.

Potassium fertilizers are used successfully in closed irrigation systems. Material costs, plant nutritional requirements, and chloride tolerance determine the appropriate source to be used. Dry materials are dissolved prior to use. Potassium solutions used in microirrigation systems should be filtered to remove solid contaminants.

Solutions containing various forms of micronutrients may be used in closed irrigation systems. The solubility of the material being used is usually more important than the form. However, synthetic chelate micronutrients are less likely to form insoluble precipitates. Insoluble components should be filtered prior to introduction into closed systems.

FOLIAR APPLICATIONS

A portion of plant nutritional needs may be met by applying fertilizer solutions directly to the foliage. *Foliar fertilization* is a useful technique in the nutritional management of many plants and has several potential benefits including:

- Supplying nutrients during periods of peak demand and when an immediate response is desired
- Supplying certain nutrients, notably micronutrients, when growing media or plant conditions are not conducive to root uptake
- Allowing precise timing of nutrient application related to the quality characteristics of the plant

- Reducing nitrate leaching in certain growing systems
- Providing a source of primary and secondary nutrients to supplement soil applications

In certain situations, such as those involving micronutrients, foliar nutrition may offer the most economical and reliable method for correcting and/or preventing deficiencies. For most plants, only a portion of the primary and secondary nutrient requirements can be supplied through the foliage. As a general rule, foliar fertilization should supplement, rather than replace, sound soil-based fertilization programs. Growers interested in foliar fertilization should seek the counsel of horticulturists, extension specialists, or agronomists regarding the plant to be treated, nutrients and sources that can be applied, and rates and methods of application.

Foliar application of nutrients may be achieved by spraying or by introducing fertilizers into overhead sprinkler systems. Spraying is usually preferred to overhead irrigation application because of more uniform distribution and less likelihood that the nutrients will be washed off before absorption occurs.

Ground spray equipment used for foliar fertilization is usually of the high-pressure, low-volume type designed to distribute the spray materials uniformly on the foliage and keep water volumes to a minimum. The spray may be applied through single or multiple-nozzle handguns; multiple-nozzle booms; multiple-nozzle, oscillating, or stationary cyclone-type sprayers; or backpack mist sprayers. Plant response may be affected by droplet size, which can be regulated by adjusting pressure and selecting the proper nozzles and discs.

Aerial or ground sprayers usually employ the same equipment used for pesticide applications. In some cases, fertilizers and pesticides may be combined. For more information, see the section on fertilizer-pesticide mixtures later in this chapter.

Plant response to foliar nutrition will depend on the growing medium, the plant species, and the environmental conditions. Specific elements differ in their rate of uptake through foliage and in the degree of mobility within the plant once absorbed (see Table 9-1). Furthermore, the chemical form of a nutrient (e.g. urea vs. nitrate) may affect responses to foliar nutrition. The form of micronutrient applied (e.g. chelate vs. sulfate) may also impact plant response.

TABLE 9-1
Generalized Absorption and Mobility Rankings for Foliar Applied Nutrients

Absorption	Mobility
Rapid	**Mobile**
Urea	Urea
Potassium	Potassium
Zinc	Phosphorus
Moderate	Sulfate
Calcium	**Partially Mobile**
Sulfate	Zinc
Phosphorus	Copper
Manganese	Manganese
Boron	Molybdenum
Slow	Boron
Magnesium	**Immobile**
Copper	Iron
Iron	Calcium
Molybdenum	Magnesium

Factors that may improve the effectiveness of foliar nutrient sprays include:

- Application during early morning or evening hours
- Application when temperatures are less than 85°F
- Relative humidity greater than 70 percent
- Wind speed less than 5 miles per hour
- Inclusion of a high-quality spreader adjuvant
- Application to young, actively growing tissue as compared to older, hardened-off tissue

If urea is used for foliar fertilization, it should have a low biuret content. *Biuret* is a compound formed during the manufacture of urea. It is toxic to some plants and can be especially damaging when applied to foliage. Ideally, biuret concentration should not exceed 0.25 percent and urea solutions should be buffered to below pH 7.0 to prevent ammonia injury.

TABLE 9-2

Target Weights for Calibration of Fertilizer Application Machinery Pounds of Fertilizer per 100 Feet of Row

Desired Rate per		Row Width in				
Acre	1,000 sq ft	18	24	30	36	48
250	5.7	0.88	1.0	1.25	1.50	2.0
500	11.5	1.25	2.0	2.50	3.50	4.5
750	17.2	2.50	3.0	3.75	4.50	7.0
1,000	23.0	3.00	4.5	5.75	7.00	9.0
1,500	34.4	5.00	6.5	8.50	10.50	14.0
2,000	45.9	6.50	9.5	11.00	13.50	18.0
2,500	57.4	8.50	11.5	14.50	17.00	23.0
3,000	68.9	10.50	14.0	17.50	21.00	28.0

CALIBRATION OF APPLICATION EQUIPMENT

The equipment manufacturer may predetermine the rate-of-delivery settings on a fertilizer applicator. Charts showing the settings or orifice sizes for rates of application according to the kinds of fertilizer are usually affixed to the equipment or are included in the operator's manual. Fertilizer dealers may be consulted for further information. Tables 9-2 to 9-4 give selected calibration information for both liquid and dry fertilizers.

The following is an example of how to use Table 9-4. To apply fertilizer at the rate of 300 pounds per acre with 400 feet of lateral moved at 60 feet setting, the pounds of fertilizer applied at each setting of the lateral is calculated as follows. Opposite a lateral length of 400 feet, find 55 pounds fertilizer to be applied per setting. Multiply 55 pounds by three to give 165 pounds to apply at each setting of the lateral for the desired 300 pounds per acre.

One can use the following equation to determine the fertilizer application rate for sprinkler systems that have sprinkler and lateral spacing other than those used in Table 9-4:

$$\text{100 lb/acre rate} = \frac{[\text{No. of Sprinklers}] \times [\text{Sprinkler Spacing (ft)}] \times [\text{Lateral Spacing (ft)}]}{435.6}$$

TABLE 9-3

Calibration of Liquid Flow

Gallons per Hour Desired	Seconds to Fill 4 oz Jar	Seconds to Fill 8 oz Jar
0.5	225	450
1	112	224
2	56	112
3	38	76
4	28	56
5	22	44
6	18	36
7	16	32
8	14	28
9	12	24
10	11	22
12	9	18
14	8	16
16	7	14
18	6	12
20	5.5	11

TABLE 9-4

Calibration of Fertilizer Rate Through a 40 ft by 60 ft Sprinkler Irrigation System

Lateral Length (Feet)	No. Sprinklers at 40 ft Spacing	Area Covered by 60 ft Setting in Acres	Quantity to Apply per Setting for Rate of 100 lb per Acre
160	4	0.22	22
240	6	0.33	33
320	8	0.44	44
400	10	0.55	55
480	12	0.66	66
560	14	0.77	77
640	16	0.88	88

Fertilizer Spreader Calibration

The following general procedure can be used to calibrate any type of fertilizer application equipment.

1. Fill spreader with a *weighed* amount of fertilizer
2. Layout calibration area in length
3. Apply the fertilizer the length of the run
4. Measure the width of the spread
5. Weigh the remaining fertilizer
6. Subtract the *remaining* fertilizer from starting amount to get applied weight
7. Calculate the applied area as:
   Length of run × width of spread = applied area in square feet
8. Divide applied weight by applied area to get lb/sq ft
9. Multiply lb/sq ft by 1,000 to get lb/1,000 sq ft or 43,560 to get lb/acre

Example: A fertilizer spreader filled with 10 pounds of fertilizer throws a pattern 7 feet wide over a 100 foot run length with 3 pounds of fertilizer remaining in the hopper at the end of the run. The spreader applied 7 pounds of fertilizer over 700 sq ft or 0.01 lb/sq ft or 10 lb/1,000 sq ft or 435.6 lb/acre.

FERTILIZER-PESTICIDE MIXTURES

Fertilizer-pesticide mixtures are used to fertilize plants and control insects, diseases, nematodes, and weeds. In a dry fertilizer-pesticide mix, the fertilizer is impregnated with the pesticide. Pesticides have been successfully combined and applied with both mixed liquid fertilizers and nitrogen solutions. Growers are encou-raged to consult with their horticulturists, extension specialists, agronomists, and pest control advisers for complete details regarding compatibility, stability, and use of fertilizer-pesticide mixtures in horticulture.

It is recommended to perform a jar test to check fertilizer compatibility with pesticides prior to application. A jar test is performed with the following precautions:

- Always wear personal protective equipment (PPE) when pouring or mixing pesticides
- Perform this test in a safe area away from food and sources of ignition

- Pesticides used in this test should be put into the spray tank when completed and applied to a labeled site
- Rinse all utensils and jars and pour the rinse water into the spray tank
- Do not use utensils or jars for any other purpose after they have contacted pesticides

The jar test procedure is performed as follows:

1. Measure 1 pint of water into a clear quart jar. Use the same water (or other diluent) that you will use when making up the larger mixture.

2. Add ingredients, stirring each time a formulation has been added, in the following order:
   - Compatibility agents and activators.
     Add 1 teaspoon for each pint per 100 gallons of final spray mixture.
   - Wettable powders and dry flowables.
     Add 1 tablespoon for each pound per 100 gallons of final spray mixture.
   - Water soluble concentrates or solutions.
     Add 1 teaspoon for each pint per 100 gallons of final spray mixture.
   - Emulsifiable concentrates.
     Add 1 teaspoon for each pint per 100 gallons of final spray mixture.
   - Soluble powders.
     Add 1 teaspoon for each pint per 100 gallons of final spray mixture.
   - Remaining adjuvants and surfactants.
     Add 1 teaspoon for each pint per 100 gallons of final spray mixture.
3. After mixing, let the solution stand for 15 minutes. Stir well and observe the results. Feel the sides of the jar to determine if the mixture is giving off heat. If so, the mixture may be

undergoing a chemical reaction and the pesticides should not be combined. Let the mixture stand for about 15 minutes and feel again for unusual heat. If scum forms on the surface, if the mixture clumps, or if any solids settle to the bottom (except for wettable powders), the mixture probably is not compatible. Finally, if no signs of incompatibility appear, test the mixture on a small area of the surface where it is to be applied.

Even if compatibility and stability are not a problem, the nature of the job to be accomplished and the grower's management practices are important considerations in making the decision to use fertilizer-pesticide mixtures. For example, it would not be wise to apply an herbicide with a phosphate fertilizer when the herbicide must be lightly incorporated, since the phosphate fertilizer should be incorporated deeper for best efficiency.

Apart from the agronomic standpoint, there is a regulatory aspect to the use of fertilizer-pesticide mixtures. Legislation, for example, provides that only qualified, licensed individuals can formulate, sell, recommend, or apply pesticides. Once a pesticide is added to a fertilizer, the mixture takes on a completely different set of legal considerations regarding its use. Also application rates are usually dependent upon the pesticide application rate.

There are a number of fertilizer-pesticide products available. It is important to read and follow label instructions for specific uses.

SUPPLEMENTARY READING

Belzer, P. 1994. Point Injection: Viable Option for Growers. *Fluid Journal Winter* 1994. pp. 14–16.

Burt, C., K.O. O'Connor, and T. Ruehr. 1995. *Fertigation.* Irrigation Training and Research Center, California Polytechnic State University, San Luis Obispo.

Petroff, Reeves. *Pesticides Interactions and Compatibility*. Montana State University. www.pesticides.montana.edu/PcideProfiles/interactions_compatibility.pdf.

Schroeder, C.B., et al. 1997. *Introduction to Horticulture: Science and Technology,* 2nd ed. Interstate Publishers Inc., Danville, IL.

Tisdale, S.L., W.L. Nelson, J.D. Beaton, and J.L. Havlin. 1993. *Soil Fertility and Fertilizer,* 5th ed. The Macmillan Company, New York, NY.

Waddington, D.V., R.N. Carrow, and R.C. Shearman. 1992. *Turfgrass.* Agronomy Monograph 32. American Society of Agronomy, Madison, WI.

CHAPTER 10

Hydroponics

The term *hydroponics*, formed by combining two Greek words—*hydro* meaning water and *ponos* meaning work—refers to the growing of plants without the use of soil. Hydroponics culture typically has plant roots periodically bathed in a nutrient solution containing all or most of 14 essential plant mineral elements: nitrogen (N), phosphorus (P), potassium (K), calcium (Ca), magnesium (Mg), sulfur (S), boron (B), chlorine (Cl), copper (Cu), iron (Fe), manganese (Mn), molybdenum (Mo), nickel (Ni), and zinc (Zn). For more information on plant essential nutrients, see Chapter 2. The formulation of the nutrient solution provides the essential plant mineral elements in proportion to plant requirements.

This chapter is an introduction to hydroponics and not an exhaustive review. For more detailed information, refer to the Supplementary Readings at the end of the chapter.

Hydroponic growing techniques are divided into two major categories:

- Nutrient solution techniques where plant roots are bathed with nutrient solution without a rooting medium
- Solid substrate techniques where plants are rooted in a solid substrate periodically bathed by a nutrient solution

In addition, there are two sub-categories based on how the nutrient solution is handled:

- Closed system in which the nutrient solution is recovered and recirculated
- Open system in which the nutrient solution is discarded after bathing the plant roots

With all hydroponic growing systems, the plants must be rooted seedlings, the stage of plant growth being such that plants can immediately grow in the chosen hydroponic rooting environment. Plants from seeds must be germinated in either a solid matrix, such as a rockwool cube, or in a matrix in which the plant roots can be easily removed. The seedling roots are placed into either a medium-free hydroponic system or a rooting medium.

NUTRIENT SOLUTION TECHNIQUES

There are three hydroponic growing techniques in which the plant is grown without the use of a rooting medium.

Standing Aerated or Circulating Nutrient Solution

This method of growing plants without rooting medium has plant roots suspended in a continuously aerated nutrient solution. This method is commonly used for conducting plant nutrition research, since the elemental composition of the nutrient solution can be easily controlled. There are no set number of plants, volume of nutrient solution, or elemental concentration parameters for this hydroponic method. In general, the greater the volume of nutrient solution per plant, the lower its elemental content can be and the less frequently the nutrient solution will have to be renewed or replaced with fresh solution. Plant species, rate of growth, and stage of growth are also factors that will determine the volume of nutrient solution per plant. Plants will grow best when the volume of nutrient solution per plant is large with a low nutrient solution elemental concentration.

There is a commercial application of this technique suited for flower and herb production. The plants are placed in raft openings and the rafts float on a pool of circulating, aerated nutrient solution

FIGURE 10-1. Plants grown on rafts floated on circulating pools of nutrient solution.

(Figure 10-1). The volume and depth of the nutrient solution, circulation rate, and elemental content are designed to be adequate to grow a plant from seedling to harvest without replenishing the nutrient solution. The nutrient solution pool depth will be a factor in maintaining the desired temperature for best plant growth.

Nutrient Film Technique (NFT)

With this technique, plant roots are suspended in a channel or trough in which the plant roots are intermittently bathed in a flow of nutrient solution, the slope of the channel determining the nutrient solution flow rate. With long channel lengths, the growth of the plants at the end of the channel may be adversely affected due to reduced aeration. Channels can be made in varying widths and depths using various types of materials, even folded plastic sheeting. For long-season plants, roots will fill the channel resulting in an uneven flow of nutrient solution, with the development of anaerobic conditions that will eventually result in root death. When growing long-season plants, deep and wide channels are needed to accommodate the large root mass. However, the NFT method is best suited for short growing period plants, such as flowering annuals and herbs (Figure 10-2). The NFT method is a closed system where

FIGURE 10-2. Plants grown on sloping NFT troughs.

water is added to maintain the solution at its original volume. The solution is filtered, sterilized, and elemental concentrations are readjusted before being recirculated.

Aeroponics

Aeroponics is a hydroponic method where plant roots are suspended in an enclosed chamber and sprayed either continuously or intermittently with a nutrient solution. Spray droplet size is determined by the size of the nozzle openings—the smaller the openings, the finer the droplet size but the greater the pressure needed to discharge the nutrient solution. The nutrient solution must be particle-free so as not to clog the nozzle openings. The finer the droplet size, the greater will be the retention of nutrient solution on the roots. Plants under high evapotranspiration demand may wilt unless another source of available water is provided. In such situations, plant roots are allowed to drop into a reservoir of nutrient solution in the base of the enclosed root chamber. Aeroponics is best suited for small plants and herbs where the roots are the harvested plant part. The aeroponic technique has been found to be an effective method for rooting woody stems, particularly for those plants that are difficult to root. Aeroponic systems are typically closed, where the nutrient solution is recirculated. Water is added to maintain its original volume, filtered to remove suspended particles, sterilized to reduce the incidence of disease, and necessary elements are replaced before recirculation.

SOLID SUBSTRATE TECHNIQUES

There are three hydroponic growing techniques in which the plants are rooted in either an inorganic or organic substrate or a mixture of each. Examples of common inorganic substrates are: pea gravel, sand, volcanic rock, perlite, or rockwool. Common organic substrates are milled and composted pine bark, sawdust, sphagnum peat, or coir. A mixture of both inorganic and organic substances can also be used as a rooting medium. The selection of the rooting medium is determined by the requirements of the hydroponic technique and the rooting physio-chemical properties desired. Each of these rooting medium substrates have their own unique volume, weight, water-holding capacity, drainage characteristics, aeration properties, cation exchange capacity, and long-term physical stability. Therefore, not all are suitable for use with each of the following hydroponic growing systems.

Flood-and-Drain or Ebb-and-Flow

This solid substrate hydroponic method consists of plants grown in a rooting medium that is periodically flooded with nutrient solution, which is then allowed to drain back into a storage vessel. The frequency of flooding is determined by the evapotranspirative demand of the growing plant. Plants are grown in either pea gravel, coarse sand, or volcanic rock in a water-tight container. The depth of the rooting medium and size of the container will be determined by the design characteristics of the entire growing system and plants to be grown. Water retention will be determined by the particle size and drainage property of the rooting medium.

This method is prone to root disease, and with time, the accumulation of unused nutrient elements retained in the rooting

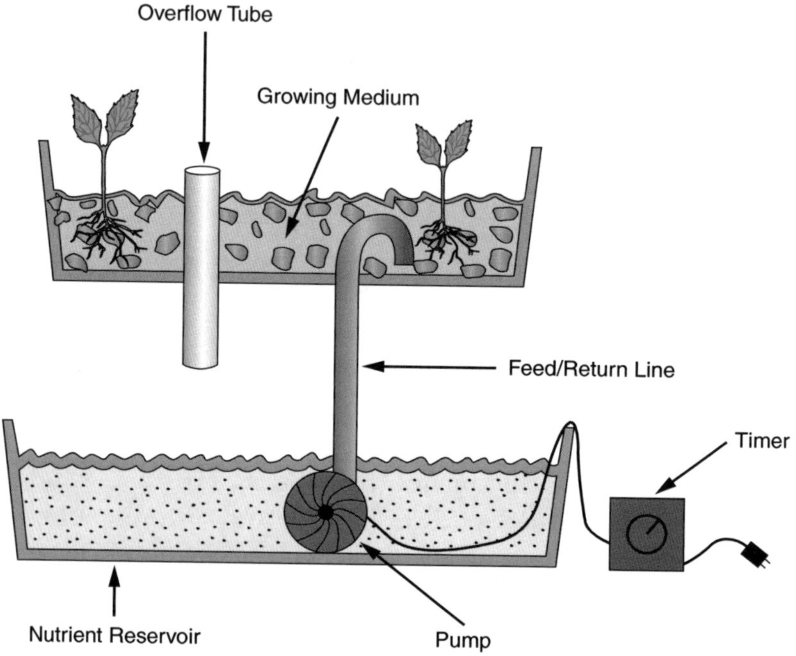

FIGURE 10-3. A flood-and-drain hydroponic growing system.

medium will begin to adversely affect plant growth. Elemental accumulation can be minimized by periodically leaching the rooting medium with water. Recirculation of the nutrient solution will require volume adjustment by adding water, filtering to remove suspended particles, sterilization to reduce the incidence of disease, and elemental reconstitution. However, with repeated use, eventually the nutrient solution will require replacement. This method of hydroponic growing was in wide use by the Army in World War II to produce fresh vegetables for troops operating in the Pacific area, and then in fairly wide commercial use for the production of tomatoes and cucumbers in tropical and subtropical areas into the 1980s. Today it is primarily used for small hobby-type growing systems (Figure 10-3).

Drip Irrigation

This method of hydroponic growing is currently in wide use for the production of many different plants in a variety of configurations.

FIGURE 10-4. A drip irrigation system with roses.

Some examples of rooting substrates used in pots and towers are inorganic, organic, or mixtures of both substrates. Another method may use rockwool or coir slabs as the rooting medium. The nutrient solution is delivered to the medium-containing pot or tower, or to the base of an individual plant using a drip irrigation system (Figure 10-4). The scheduling of nutrient solution delivery is based on the water requirement of the plant, and frequency and volume based on stage of plant growth. Drip systems are normally operated as an "open" system, where nutrient solution is applied until there is a slight overflow and the overflow is discarded with each nutrient solution application. Depending on the rooting medium, nutrient solution composition, and volume delivered, there occurs a retention and accumulation of unused nutrient solution in the rooting medium. The retained nutrient solution requires electrical conductivity (EC) monitoring, and when reaching a certain EC level, the rooting medium requires leaching with clean water. This leaching requirement is costly in terms of both water used and the need to properly discard the leachate. Plant species, stage of growth, and frequency and volume of nutrient solution delivered with each irrigation are the factors that will determine the elemental content of the delivered nutrient solution.

Sub-Irrigation

In this technique, the nutrient solution is introduced at the base of the rooting medium at a sufficient volume to maintain a constant depth of about one inch of nutrient solution. Most systems of this type use an automatic means for maintaining a constant level of nutrient solution. The sub-irrigation technique is the most efficient of all hydroponic growing methods as all of the applied water and nutrient elements are utilized by the plant. The rooting medium selected must have a sufficient wicking characteristic so that the upper portion of the rooting medium will be moist, but not water logged, so as to keep the plant roots actively functioning. The depth of the rooting medium will be determined, in part, by its wicking property. The elemental content of the nutrient solution must not exceed plant needs. Those rooting media that lack biological stability are not suited for use with this hydroponic method.

THE NUTRIENT SOLUTION

Elemental Content

A complete nutrient solution must contain the 14 essential plant mineral elements. The major elements found in a complete nutrient solution are nitrogen, phosphorus, potassium, calcium, magnesium, and sulfur, all of which are required at relatively high concentrations. The essential plant micronutrients are boron, chlorine, copper, iron, manganese, molybdenum, nickel, and zinc, and are required at lower concentrations than the major elements. A nutrient solution formula may not include one or more elements depending on the supply of elements present in the rooting medium or the water used.

Elemental Form

All 14 essential mineral plant elements must exist in ionic form in solution in order to be absorbed by plant roots. The ionic forms of the macronutrients and micronutrients in a nutrient solution are listed in Table 10-1.

The ionic form of phosphorus that predominates in the nutrient solution depends on solution pH. Under acid pH, dihydrogen

TABLE 10-1
Essential Plant Mineral Elements and Their Ionic Forms

Macroelements	Forms	Ions	Microelements	Forms	Ions
Nitrogen			Boron		BO_3^{3-}
	Ammonium	NH_4^+	Chloride		Cl^-
	Nitrate	NO_3^-	Copper		Cu^{2+}
Phosphorus			Iron		
	Trihydrogen	H_3PO_4		Ferrous	Fe^{2+}
	Dihydrogen	$H_2PO_4^-$		Ferric	Fe^{3+}
	Monohydrogen	HPO_4^{2-}	Manganese		Mn^{2+}
Potassium		K^+	Molybdenum		MoO^{3-}
Calcium		Ca^{2+}	Nickel		Ni^{2+}
Magnesium		Mg^{2+}	Zinc		Zn^{2+}
Sulfate		SO_4^{2-}			

phosphate ($H_2PO_4^-$) or monohydrate phosphate (HPO_4^{2-}) predominate. When the pH is near neutral or alkaline, tri-phosphate (PO_4^{3-}) predominates in nutrient solutions.

Two ions, the nitrate (NO_3^-) anion and potassium (K^+) cation, are easily absorbed by plant roots and are present in most nutrient solutions in fairly high concentrations. All the other element ions in the nutrient solution are more selectively absorbed by plant roots.

In nutrient solution formulations, the need for the chelated forms of the micronutrients, particularly iron, but also copper, manganese, and zinc, is questionable. In one study, it was demonstrated that the chelate, ethylene-diamine-tetraacetic acid (EDTA), could be toxic to plants when added at high concentrations in nutrient solutions. Therefore, some nutrient formulations use other chelates, such as diethylene-triamine-pentaacetic acid (DTPA), that is not thought to be toxic to plants. Although chelated forms for these micronutrients have proven to be of value based on certain soil conditions—particularly in alkaline and organic soils, and organic soilless rooting media—their use in hydroponic nutrient solutions may not be justified in terms of improved availability. When adding a chelate to a mix of elements in solution, the stability of the initial chelate will depend on the concentration of the other ions in solution as well as the nutrient solution pH. Chelation can be significantly reduced under certain situations, which negates the reason that the chelated form of the element was selected over non-chelated elemental forms. The following inorganic forms of iron will keep iron in solution, and therefore, be able to meet the Fe requirement of the plant:

- Iron (ferrous) sulfate ($FeSO_4 \cdot 7H_2O$)
- Iron (ferric) sulfate [$Fe_2(SO_4)_3$]
- Iron (ferric) chloride ($FeCl_3 \cdot 6H_2O$)
- Iron ammonium sulfate [$(NH_4)_2SO_4 \cdot FeSO_4 \cdot 6H_2O$]

Concentration Range and Ratios

All 14 essential mineral plant elements in a nutrient solution must be within a particular concentration range, as well as in a particular ratio among certain elements in order to ensure sufficiency. Most nutrient solution formulations contain higher concentrations of some elements than required by the plant. Phosphorus and nitrogen fall into this category. In some nutrient solution formulations, the ratios among the major cation elements, potassium (K^+), calcium (Ca^{2+}), and magnesium (Mg^{2+}), are frequently out of balance.

Elements in a nutrient solution interact among themselves, exhibiting both antagonistic and synergistic characteristics. For example, among the major cations potassium, calcium, and magnesium, the least competitive is magnesium. Magnesium deficiency is likely to occur with the use of some nutrient solution formulations having high concentrations of potassium or calcium when growing magnesium-sensitive plants. The ammonium (NH_4^+) cation is a strong competitor and can reduce the uptake of both calcium and magnesium. There is a synergistic relationship between NO_3–N and potassium. The presence of high concentrations of NO_3–N enhances the uptake of potassium. The presence of a low concentration of ammonium in a nutrient solution will enhance the uptake of NO_3–N.

Water Quality

Water quality is an important factor to consider in hydroponic culture. The formulation of a nutrient solution requires the use of relatively pure water. Water must be free from those substances than can be potentially toxic to plants. A water quality analysis by a reliable laboratory should be performed to determine if the water selected is suitable for hydroponic use. Reverse osmosis is the best method for removing unwanted substances from water.

Formulation

A nutrient solution is made by dissolving element-containing sources in water to achieve a certain elemental concentration. Most nutrient solution formulations are based on the two formulations published by Hoagland and Arnon in 1950.

A common means for formulating a nutrient solution is to prepare concentrates of certain groups of elements, known as *stock solutions*. Since several essential plant elements are incompatible when mixed in a concentrated form, such as calcium and phosphate, two or three separate stock solutions are needed. Each stock solution is frequently designated by a letter, such A, B, and C. A portion of the concentrated stock solution is injected into a flowing water stream to obtain the desired elemental concentration in the final nutrient solution. In addition, if pH correction is required, acids such as sulfuric (H_2SO_4), phosphoric (H_3PO_4), or nitric acid (HNO_3), or bases such as potassium hydroxide (KOH), will be injected into the flowing stream of water.

Fertilizers for preparing stock solutions are typically of higher purity than the same fertilizer-grade material used in field grown situations. The fertilizers must be water soluble and should be relatively free of impurities. Some of the common major element sources are magnesium sulfate or Epsom salts, calcium nitrate, monopotassium phosphate, ammonium nitrate, and potassium nitrate. Micronutrient sources for copper, iron, manganese, and zinc are either their sulfate or chloride salts, or as a chelate.

An example of a potential pair of stock solutions for formulating a nutrient solution is given in Table 10-2.

TABLE 10-2

Example for a Pair of Stock Solutions

Source	50 Gallons	10 Liters
Tank A		
Potassium Nitrate	21 lb	503 g
Monopotassium Phosphate	12 lb	288 g
Epsom Salts	21 lb	503 g
Boric Acid	54 g	2.8 g
Manganese Sulfate	28 g	1.5 g
Zinc Sulfate	4 g	0.2 g
Copper Sulfate	1 g	0.05 g
Molybdic Acid	0.5 g	0.03 g
Tank B		
Calcium Nitrate	45 lb	1,079 g
Iron EDDHA 6% Fe	2 lb	48 g

Diluting the Tank A and B stock solutions at a ratio of 1:200 (e.g. 2 quarts in 100 gallons of water) gives a final concentration of elements as shown in Table 10-3.

Plant elemental requirements differ among plant species. The final nutrient solution concentrations must match the element requirements of the plants being grown. Frequently checking the electric conductivity (EC) of the formulated nutrient solution is one way to ensure that a plant is receiving an optimal concentration of its required elements. If the EC is either greater or less than the

TABLE 10-3

Element Concentrations After 1:200 Dilution of Stock Solutions from Tanks A and B

	N	P	K	Ca	Mg	S	B	Cu	Fe	Mn	Mo	Zn
ppm*	119	30	140	100	24	32	0.25	0.01	3.0	0.25	0.005	0.025
meq/L**	8.5	1	3.5	5	2	2						

*parts per million ; **milliequivalents per liter

recommended value, the injection rate of stock solutions should be decreased or increased, respectively. Some suggested values for the EC and nutrient elemental concentrations for several plants being grown hydroponically are given in Table 10-4.

Some plant essential elements are taken up at faster rates than others. This explains why some of these plant elements are in nutrient solutions at concentrations greater than the theoretical plant demand. If elements such as nitrogen and phosphorus were applied at concentrations that matched plant demand, the nutrient solution would quickly become depleted of these elements upsetting the critical balance between all the plant required elements. This becomes very important when the nutrient solution is recirculated. The relative element absorption rates by plant roots are given in Table 10-5.

TABLE 10-4

Suggested ECs* and Element Concentrations (ppm) of Nutrient Solutions for Four Plant Species**

	Carnation	Rose	Daisy	Chrysanthemum
EC dS/m	1.7	1.5	1.5	1.4
NO_3–N	182	154	158	147
NH_4–N	14	18	21	14
P	38	38	38	31
K	244	196	215	196
S	40	40	40	32
Ca	150	140	120	110
Mg	24	18	24	24
Fe	1.4	1.4	2.0	3.4
Mn	0.5	0.3	0.3	1.1
Zn	0.3	0.2	0.3	0.2
B	0.3	0.2	0.3	0.2
Cu	0.05	0.05	0.05	0.03
Mo	0.05	0.05	0.05	0.05

*electrical conductivity; **parts per million

Sonneveld, C., and N. Staver. 1990. *Naaldwijk Research Station for Plants Under Glass.* Publication No. 8.

TABLE 10-5

Relative Element Absorption Rates by Plant Roots from a Hydroponic Nutrient Solution

Fast Active	NO_3, NH_4, P, K, Mn
Intermediate	Mg, S, Fe, Zn, Cu, Mo
Slow Passive	Ca, B

www.usu.edu/cpl/research_hydroponics3.htm#silicon.

Beneficial Elements

There are several elements that have been identified as being beneficial to plants, but not essential. Some of these elements are natural contaminates that may be present in the rooting medium, fertilizers, and water used to formulate the nutrient solution. Of all the nonessential elements, silicon (Si) is probably the only element that might be beneficial. Silicon has been found to give some plants resistance to insect and disease infestation if included in the nutrient solution formulation. Therefore, the addition of a source of silicon to the nutrient solution, such as potassium silicate, may be desirable.

Recirculated Nutrient Solution Monitoring

Differential ion uptake occurs with each use of a nutrient solution. A laboratory analysis may be required to determine what changes in elemental concentrations have occurred in the nutrient solution in order to bring it back to its initial formulation. The pH of a recirculated nutrient solution may also require adjustment using acids, such as sulfuric, nitric, or phosphoric to reduce the pH, and potassium hydroxide to increase the pH.

The main advantages for recirculating a nutrient solution are savings in fertilizer and water. One disadvantage of recirculating nutrient solutions is the potential for rapid build-up of root diseases, such as *Pythium sp.* Disease control requires both filtering and sterilization of the nutrient solution, using either UV radiation or ozone. Some recommend bubbling either air or oxygen into the nutrient solution before recirculation.

Other Monitoring Tools

Plant or tissue analysis is a tool for monitoring the elemental status of the growing plant. General guidelines for plant tissue sampling and critical element levels corresponding to plant species are described in Chapter 11.

Consultation with Cooperative Extension Service representatives, other university personnel, private consultants, or industry experts can be of valuable assistance to the hydroponics grower.

SUPPLEMENTARY READING

Biondo, R.J., and J.S. Lee. 1997. *Introduction to Plant and Soil Science and Technology.* Ronald J. Interstate Publishers, Inc., Danville, IL.

Bugbee, B. 2003. Nutrient management in recirculating hydroponic culture. South Pacific Soilless Culture Conference, Feb. 11, 2003, Palmerston North, New Zealand.

Copper, A. 1996. *The ABC of NFT, Nutrient Film Technique.* Casper Publications, Narrabeen, Australia.

De Kreij, C., C. Sonneveld, and M.G. Warmenhoven. 1987. *Guide Values for Nutrient Element Contents of Vegetables and Flowers Under Glass.* No. 15. Glasshouse Crops Research and Experiment Station, Naaldwijk, The Netherlands.

Hoagland, D.R., and D.I. Arnon. 1950. *The Water Culture Method for Growing Plants Without Soil.* Circular 347: 1–32. Agricultural Experiment Station, University of California, Berkeley, CA.

Jones, Jr., J. Benton. 2005. *Hydroponics: A Practical Guide for the Soilless Grower.* 2nd Edition. CRC Press, Boca Raton, FL.

Resh, H.M. 2001. *Hydroponic Food Production*, 6th Edition. New Concept Press, Mahwah, NJ.

Schroeder, C.B., et al. 1997. *Introduction to Horticulture: Science and Technology,* Second Edition. Interstate Publishers, Inc.

Sonneveld, C. 1985. *A Method for Calculating the Composition of Nutrient Solutions for Soilless Cultures,* Second Translated Edition. No. 10. Glasshouse Crops Research and Experiment Station, Naaldwijk, The Netherlands.

Yuste, M.P., and J. Gostinear (eds). 1999. *Handbook of Agriculture*. CRC Press, Boca Raton, FL.

CHAPTER 11

Soil, Media, and Tissue Testing

Soil, soilless media, and plant tissue analyses are among the best guides for the sensible and efficient use of fertilizers and soil amendments. Laboratory analyses are an integral part of fertilizer best management practices (BMPs) that aid in the production of vigorous, healthy plants while maintaining environmental quality. For more information on BMPs see Chapter 12. Useful recommendations resulting from soil, media, and plant tissue tests require correct sampling, accurate analysis, and interpretation based on sound research, practical experience, and good judgment.

Soil, media, and plant tissue analyses have strengths and limitations. Different information is gathered from each analytical method. Testing should be used so that the results from different methods support and supplement each other. For example, soil and media analyses are most useful for appraising general nutrient status, evaluating pH, and salinity. With this information, soil or media can be supplemented prior to planting or during the growing season to provide balanced plant nutrition.

Plant tissue analysis consists of analyzing leaves or other plant parts to determine what nutrients are in the plant at the time of sampling. Results from tissue analyses give an indication of what plants are absorbing from the root zone. Plant tissue analyses are particularly useful for monitoring the nutritional status of greenhouse and nursery plants and turf. Plant analysis is also useful for diagnosing problems of trees and shrubs in the landscape where soil samples of the entire root zone are difficult to obtain and interpret. Tissue analyses are also useful for diagnosing the potential causes of poor growth, evaluating the effectiveness of fertilizer treatments, monitoring the nutrient status of plants throughout the growing season, and managing plant quality.

Satisfactory recommendations based on soil, media, or plant tissue tests depend upon three factors: collection of a representative sample, accurate analysis, and proper interpretation of the analytical results.

The terms "soil" and "media" are used throughout this chapter. Soil occurs naturally in the landscape. Soilless media is "man-made" and is commonly used in greenhouse and container nursery operations. The differences in terms of analytical methods and interpretation between soil and media will be discussed in more detail later in the chapter. For more in-depth information, refer to Chapter 3 Soils and Chapter 4 Soilless Media.

SOIL AND SOILLESS MEDIA SAMPLE COLLECTION

The first step in soil and media testing is collecting a representative sample. The sample must accurately represent the soil or media in which plants are growing or are to be grown. Proper sampling is important because the results obtained from the laboratory can be no better than the sample submitted. A sample submitted to a laboratory will usually weigh less than a pound, but may represent millions of pounds of soil or many cubic yards of media. Fertilizer and amendment recommendations will be based on the results of the laboratory analysis. Improper sample collection can lead to incorrect recommendations.

Soil Sampling

Soils in the landscape differ from area to area. Soils may vary in chemical and physical characteristics in both their horizontal and vertical dimensions (Figure 11-1). Management practices such as grading, fertilizing, and topdressing can further increase physical and chemical differences. Soil variability should be recognized and taken into account when samples are collected and when fertilizer or amendment applications are made.

Soils that vary in color or texture are probably chemically different from one another. Another indicator of varying soil characteristics are different native plants growing in different locations. Soil samples taken from different areas of the landscape that look similar may still differ chemically.

Soils from different management zones should not be mixed together. Mixing soil samples is acceptable when the location being sampled will be managed as a whole. Where practical, dissimilar soil types should be sampled and analyzed separately. As an example, soil samples taken from a high maintenance putting green that has been heavily irrigated and fertilized should not be mixed with soil taken from a nearby ornamental landscape that has a different irrigation and fertilization program. However, one may mix soil samples taken from fairways that will receive the same fertilizer program.

FIGURE 11-1. **Soil profile showing differing layers with depth.**

In a uniformly managed soil area, sub-samples should be collected from a number of different locations within the area. Sub-samples should be thoroughly mixed and a representative composite sample is sent to the laboratory for analysis. An example of a composite sampling strategy would be sampling a uniform soil from a baseball field. A convenient number of sub-samples would be nine, one from near each defensive player's location. The nine sub-samples are mixed thoroughly and one representative sample is removed for analysis. The larger the space to be sampled, the greater the possibility for variability within that area. Consequently, larger areas that are uniformly managed will require more sub-samples.

In most cases, sampling depth should be within the root zone. Turf, annuals, and many species of herbaceous perennials have shallow root systems, so a sample would be taken from approximately 1 to 6 inches in depth (Figure 11-2). Scrape away the surface residue before sampling. For deep-rooted trees and shrubs, more than one sampling depth may be needed. When sampling to identify the concentration of mobile elements, such as nitrate-nitrogen, sulfate-sulfur, sodium, or chloride, multiple depths are recommended. For example, three samples could be taken at 12 inch intervals from the

FIGURE 11-2. A soil sample taken from the root zone of a golf green.

surface to a 36 inch depth. Sub-samples taken at a specific depth must not be mixed with sub-samples taken from another depth.

Probes especially designed for soil sampling, a garden trowel, spade, or a shovel may be used to take soil samples. Soil sampling probes are made of stainless steel or are chrome plated. Other metal tools can be used when stainless steel is not available. An inexpensive plastic trowel works well when one is taking a slice from the side of a pit or for soil that is easy to dig. Tools must be cleaned before and after sampling a new soil type.

When testing a soil for plant nutrients, it is best to send samples to the laboratory on the day of sampling. If samples for nutrient testing cannot be sent to the lab immediately, they can be either air-dried at room temperature or refrigerated until they can be shipped to the laboratory for analysis. Do not allow the soil sample temperature to exceed 100°F as it can alter analytical results.

When testing soil for biological parameters, such as microorganism identification, pathology, or nematode identification, samples should be sent to the lab as soon as possible following sampling. The soils should not be frozen, allowed to dry out, or become hot. Samples should be kept moist and at room temperature.

Soil analysis is helpful when attempting to diagnose production or plant quality problems. Where problems exist, separate samples for testing should be collected from normal and problem areas. In this way, results from the normal and problem areas can be compared to determine the cause of the problem.

It is useful to obtain specific information from the testing laboratory before collecting and submitting soil samples. Determine what test or suite of tests the lab should run on your sample. Many labs have different report formats and may have an option for a written interpretation of the results. Some labs also have technical service providers or Certified Crop Advisers (CCA) available for consultation and written reports. A menu of laboratory services may be obtained from the laboratory's website or by contacting the lab. Most laboratories provide sample bags, submittal forms, and information sheets. To make procedures easier, plastic zip lock bags work well for transporting soil samples. Two to four cups of soil per sample is adequate for most common laboratory analyses.

It is important that each sample bag is identified with as much information as possible to help the lab keep track of your sample. Information to include on the sample label includes date and location sampled, the person taking the sample, identity or location of the sample, and a contact phone number. A map showing the landscape, sample locations, and sample identification will also be useful when interpreting results. There is little value in laboratory results without knowing where the samples were taken. Indicate what type of plant is growing or intended to be grown in the soil to be tested. For the commercial nursery or greenhouse, detailed records on areas sampled, fertilizers and production practices used, and plants grown will aid interpretation and recommendation.

Media Sampling

Soilless media goes by many names including soilless mixes, potting media, growing mix, greenhouse mix, nursery media, transplant mix, liner mix, canning mix, as well as other descriptive names. Media differ based on the percentage and type of components used, as well as their intended use.

There is a substantial difference between soil and soilless media. Soil is naturally occurring in the landscape and is composed of unconsolidated mineral matter and organic residues as principal physical components. Soilless media are man-made and may be composed of materials such as ground bark, peat moss, perlite, vermiculite, sand, and other components.

Soilless media has greater uniformity than soil and this simplifies sample collection. Bulk media can be sampled prior to use to identify any constituents that may have adverse effects on plant growth. Samples can also be taken during the growing period to help confirm the growing media's nutrient composition. A well-mixed growing medium can be sampled using a standard probe. A representative number of sub-samples should be collected from a pile of soilless media and mixed together to form one sample for laboratory analysis.

A soil probe can also be used to take media samples from nursery containers while plants are being grown. Take one sub-sample from several containers to create a representative sample. A similar technique can be used in the greenhouse. At least

1 gallon of media must be sent to the laboratory if testing for porosity and other physical parameters. Otherwise, a quart of media should be sufficient to run a general nutrient analysis. Check with your lab to determine the amount needed for your required testing.

SAMPLE ANALYSIS

Soil and media may be analyzed for a wide range of chemical and physical characteristics. A typical suite of tests could include: nitrate-nitrogen, ammonium-nitrogen, phosphorus, potassium, calcium, magnesium, sodium, sulfate-sulfur, zinc, manganese, iron, copper, boron, pH, and soluble salts. For an extra fee, tests for chloride, molybdenum, free lime, and organic matter may also be performed. Many laboratories offer comprehensive test packages.

Laboratory procedures for analyzing soil differ from soilless media. The volume of sample needed for a media analysis will be larger than the volume needed for a soil sample. A minimum of one quart of media is suggested for a full nutrient analysis as compared to one pint of soil.

Ideally, all soil testing laboratories would use the same procedures for testing soil or media. Although most laboratories use published methodology, actual methods used vary. Analytical methods have been developed that correlate the probability of plant growth with added nutrients. Under different conditions, different analytical methods have been found to correlate more closely with plant growth. For example, the Olsen sodium bicarbonate method is superior for predicting plant response to phosphorus in alkaline (pH > 7.0) soil, whereas the Bray method is superior in acidic soils. Both methods predict phosphorus responses well between pH 5.5 and 7.0.

It is important to use methods appropriate for your application. If you are uncertain which method is correct, consult the laboratory for its recommendation prior to sample submission. It is also important to understand that results from one analytical method cannot be directly compared with the results of a different method. One may compare results only if the same method was used in both cases. Most labs provide interpretation guidelines that give an indication of the meaning of test results.

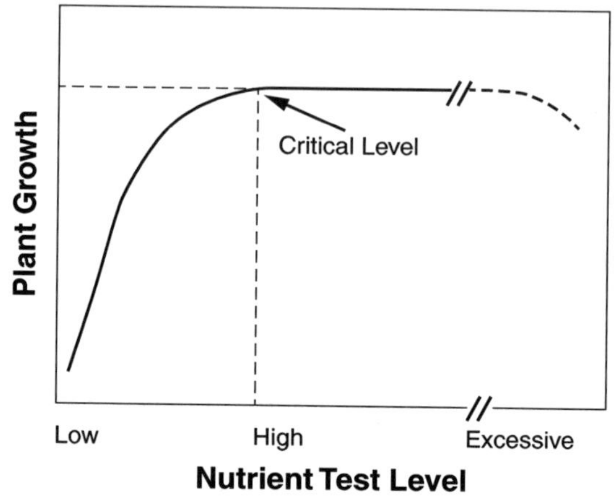

FIGURE 11-3. A typical plant growth response curve to nutrient levels.

A generalized plant response curve relating soil test values to plant growth is shown in Figure 11-3. Plant growth is limited at low soil test values of a particular plant nutrient in the soil. The *soil test critical level* is the point where increases in plant growth cease in response to further increases in the concentration of a particular soil nutrient. Sites with this soil test value or higher will usually not respond or respond adversely to an application of that nutrient. On the other hand, sites with soil test values below the critical level are more likely to respond positively to an application of that nutrient. The point on the far right where the curve turns down indicates where the nutrient is present in excessive amounts leading to reduced growth due to toxicity.

INTERPRETATION OF ANALYTICAL RESULTS

Proper interpretation of analytical results is important to the success of a soil or media testing program. Soil and media testing do not directly measure plant available nutrients. The most common techniques measure extractable nutrients and provide an index of plant nutrient availability. Soil testing provides values that are correlated with the likelihood of a response to the addition of nutrients (Table 11-1). Media testing, on the other hand, measures water-soluble or DTPA extractable nutrients (Table 11-2, page 273) and likewise provides indexes of plant nutrient availability.

The results from laboratory analysis must be interpreted before meaningful nutrient or amendment recommendations can be made. The interpretation of the results from soil and media testing

TABLE 11-1

Generalized Soil Test Interpretative Guide

Analyte *(lab method)*	**UNITS**	**LOW[1]**	**NORMAL[2]**	**HIGH[3]**
Organic Matter (OM) *(by combustion)*	Percent	<1.0	1.0-3.0	>3.0
pH *(saturated paste)*	pH	<6.2	6.2-7.8	>7.8
Soluble Salts (ECe) *(saturated paste extract)*	dS/m	<0.7	0.7-2.0	>2.0
Free Lime (AEC) *(alkaline earth carbonates)*	Percent	<2.0	2.0-3.0	>3.0
Phosphate-Phosphorus (PO_4–P) *(Bray extraction)* *(soils at pH less than 6.0)*	ppm	<30.0	30.0-50.0	>50.0
Phosphate-Phosphorus (PO_4–P) *(sodium bicarbonate)*	ppm	<12.0	12.0-25.0	>25.0
Potassium (K) *(ammonium acetate)*	ppm	<100	100-300	>300
Calcium (Ca) *(saturated paste extract)*	meq/L[4]	<5 (T)	5-30 Ca should be greater than Na and/or Mg	>30
Magnesium (Mg) *(saturated paste extract)*	meq/L[4]	(T)	1-10	Should not exceed Ca

(continued)

TABLE 11-1 (Continued)

Analyte *(lab method)*	**UNITS**	**LOW[1]**	**NORMAL[2]**	**HIGH[3]**
Sodium (Na) *(saturated paste extract)*	meq/L[4]	N.A.	<3	>5 Should not exceed Ca + Mg
Chloride (Cl) *(saturated paste extract)*	ppm	<170	170-350	>350
	meq/L	<5.0	5.0-10.0	>10.0
Nitrate-Nitrogen (NO_3–N) *(potassium chloride)*	ppm	<10.0	10.0-25.0	>25.0
Sulfate-Sulfur (SO_4–S) *(ammonium acetate)*	ppm	<10.0	10.0-15.0	>15.0
Boron (B) (T)				
(barium chloride-hot water)	ppm	<0.5	0.5-1.2	>1.2
(saturated paste extract)	ppm	<0.2	0.2-0.7	>0.7
Iron (Fe) *(DTPA)*	ppm	<5.0	5.0-15.0	>15.0
Manganese (Mn) *(DTPA)*	ppm	<1.0	1.0-10.0	>10.0
Copper (Cu) *(DTPA)*	ppm	<0.5	0.5-1.2	>1.2
Zinc (Zn) *(DTPA)*	ppm	<0.8	0.8-1.5	>1.5
Molybdenum (Mo) *(DTPA)*	ppm	<0.1	0.1-0.2	>0.2

[1] "Low" nutrient range may be inadequate supply of nutrient.

[2] "Normal" range should provide a safe and nutrient adequate growing environment.

[3] "High" nutrient range may be environmentally and economically unwise. For salts, signs of toxicity may be expected.

[4] Used to calculate ESP.

(T) Tissue analysis is often a better measure of plant adequacy for this nutrient.

N.A. Not Applicable.

TABLE 11-2

Nursery and Greenhouse Media Analysis Generalized Guidelines for Modified (DTPA) Saturated Media Extract[1]

Analyte	Description	Unit	Low	Acceptable	Optimum	High	V High
pH	pH	pH	5.0	5.5	6.8	7.5	9.0
EC_e	Conductivity	dS/m	0.5	1.0	2.0	5.0	7.0
NH_4–N	Ammonium-N	ppm	10	50	100	150	200
NO_3–N	Nitrate-N	ppm	40	70	200	300	400
PO_4–P	Phosphate-P	ppm	5	15	25	50	100
K	Potassium	ppm	50	100	150	300	400
Mg	Magnesium	ppm	20	70	150	170	200
Ca	Calcium	ppm	50	100	200	300	400
Na	Sodium	ppm	20	50	80	200	300
SO_4–S	Sulfate-S	ppm	20	70	200	300	400
Zn	Zinc	ppm	5	10	30	35	40
Mn	Manganese	ppm	5	10	30	35	40
Fe	Iron	ppm	15	25	50	75	100
Cu	Copper	ppm	1	2	10	20	30
B	Boron	ppm	0.1	0.5	1.0	2.0	3.0
Cl	Chloride	ppm	Maintain below 45 ppm				
F	Fluoride	ppm	Maintain below 5 ppm				

[1]Information obtained from commercial laboratory data and D.D. Warncke, Michigan State University. pH is determined using 1:2 ratio, media:deionized water by volume. Remaining tests using DTPA extraction.

must take into consideration such additional factors as plant species and variety, environmental concerns, irrigation, and cultural practices.

The interpretive tables presented in this chapter are general tools that help guide the user with prudent and reasonable use of fertilizers and amendments. These tables should be used as broad guidelines supplementing laboratory data obtained from samples collected and analyzed by the procedures specified. The user is advised to contact a commercial laboratory technical representative, a university cooperative extension horticultural adviser, a Certified Crop Adviser (CCA), or a manufacturer's representative for recommendations applicable to specific growing systems.

Soil nutrient balance is important when attempting to attain a productive and fertile soil at the lowest cost to the grower. To one degree or another, substances are directly or indirectly toxic or antagonistic above some threshold. For example, common nutrients such as manganese and iron may become toxic at high levels and reduce yields. The plant nutrient boron has a narrow range between deficiency and toxicity and may be toxic to some sensitive species even at a comparatively moderate soil concentration.

High levels of an exchangeable base such as sodium, potassium, or magnesium impact soil physical properties, reducing air and water movement into and through soil. Base cation ratios have been used to characterize ideal soil conditions. The values used have been: potassium 2 to 5 percent, calcium 60 to 70 percent, magnesium 10 to 20 percent, and sodium at less than 5 percent. Currently it is thought that while nutrient balance is important, the ideal basic cation saturation ratio frequently overestimates the need for amendments and fertilization. Sufficiency levels, such as those presented in Table 11-1 (page 271), are more universally applicable.

Soil test methods should not be used for soilless media. There are differences in the results obtained from a soilless media test as compared to the results for soil tests.

Soilless media analysis is beneficial in the diagnosis of production problems. Samples are collected from areas where plants are growing normally and from areas where plants are exhibiting reduced production or poor quality. Unless a specific cause is suspected, a broad range of nutrients and potentially toxic elements may need to be evaluated. The values obtained in the normal and affected areas are then compared. Be aware that a healthy plant will remove more nutrients from the media than a weakened plant. Diagnosing production problems may be complex and advice should be sought from individuals experienced in the interpretation of laboratory data.

PLANT ANALYSIS

Plant tissue testing is a tool that growers can use to improve plant nutrition monitoring. Plant structures, such as leaves, leaflets, midribs, and petioles, can be analyzed for their nutrient levels. Commonly for ornamental species and turf, the plant parts sampled and analyzed are the leaves. A plant analysis indicates what the plant is removing from the soil. Plant analysis complements soil or media testing and the two together act as tools for managing plant nutrition.

Sample Collection

Collecting a representative plant sample is essential if reliable results are to be obtained (see Figure 11-4). When collecting a plant sample, the species, the stage of growth, the plant part, and the nutrients to be measured must all be considered. Do not mix different plant species to make up one sample. For example, do not mix rose leaves with camellia leaves to make up one sample. The stage of growth will be important for growers to recognize. The concentration of some nutrients naturally declines and some nutrients increase as a plant matures. Therefore, the stage of growth is important if consistent sampling and interpretation of results are to be maintained. Nutrient concentrations can vary widely between different plant parts. Even old leaves and young leaves on the same plant can vary widely in nutrient content. For example, reproductive tissue generally has a higher phosphorus concentration than vegetative tissue. In many instances the most recently matured leaves are used for analysis. For turfgrass, clippings can be used for analysis. See Table 11-3 for specific tissue sampling recommendations for common turf and ornamental species. It is important that samples of the appropriate plant part be analyzed if the values presented in Table 11-3 are to be used correctly.

Dirty plant tissue samples can give misleading laboratory results. Dirt and dust on leaves that are processed through the laboratory may impart very high levels of manganese, iron, and

FIGURE 11-4. Most recent fully matured leaf.

aluminum that are not necessarily present within the plant tissue. This will skew some reported data of other nutrients and they will report misleadingly low. Soil commonly contains between 10,000 to 300,000 ppm aluminum. One can see how the resulting aluminum level will be elevated if soil particles contaminate plant tissue. Plant samples that have been sprayed with foliar nutrients can also give misleading results. It is best to sample plant tissue that has not received foliar fertilizer applications.

It is important to clean dirty or dusty leaves or turfgrass clippings prior to analysis. Most laboratories offer this service upon request or it is done as standard operating procedure. If the lab does not wash plant tissue, the following procedure may be used. Fill a plastic basin with water. Add a few drops of phosphate-free, liquid dish washing detergent. Place the leaves in a colander and submerge the sample in the soap and water bath. Gently move the leaves about for 8 to 10 seconds, remove them, and rinse the

TABLE 11-3

Generalized Plant Tissue Guidelines for Turf and Ornamental Species

Latin Name	Common Name	Plant Part for Sampling	Nutrient Sufficiency	Range Low to High
Turfgrasses				
Agrostis stolonifera	Creeping Bentgrass	Clippings	N	4.0 – 6.0%
			P	0.3 – 0.6%
			K	2.0 – 2.6%
			Ca	0.5 – 0.75%
			S	0.18 – 0.6%
			B	9 – 20 ppm
			Zn	25 –100 ppm
Cynodon dactylon	Common Bermu-dagrass	Clippings	N	3.0 – 5.0%
			P	0.15 – 0.5%
			K	2.0 – 4.0%
			Ca	0.25 –1.0%
			S	0.2 – 0.5%
			B	6 – 30 ppm
			Zn	25 – 200 ppm
Festuca elatior	Tall Fescue	Clippings	N	3.2 – 4.6%
			P	0.34 – 0.45%
			K	3.0 – 4.0%
			Ca	0.4 – 0.5%
			S	0.20 – 0.45%
Lolium perenne	Perennial Ryegrass	Clippings	N	4.0 – 5.0%
			P	0.35 – 0.55%
			K	2.0 – 4.0%
			Ca	0.25 – 0.6%
			S	0.27 – 0.45%
Poa pratensis	Kentucky Bluegrass	Clippings	N	2.6 – 6.0%
			P	0.28 – 0.55%
			K	2.0 – 4.0%
			Ca	0.27 – 0.6%
			S	0.18 – 0.5%

(continued)

TABLE 11-3 (Continued)

Latin Name	Common Name	Plant Part for Sampling	Nutrient Sufficiency	Range Low to High
Stenotaphrum secundatum	St. Augustine Grass	Clippings	N	2.0 – 3.0%
			P	0.2 – 0.5%
			K	2.5 – 4.0%
			Ca	0.3 – 0.5%
			S	0.2 – 0.6%
Zoysia japonica	Zoysiagrass	Clippings	N	2.0 – 2.5%
			P	0.2 – 0.4%
			K	1.0 – 1.5%
			Ca	0.3 – 0.5%
			S	0.25 – 0.45%
	Turfgrass (generic)	Clippings	N	2.75 – 4.50%
			S	0.20 – 0.45%
			P	0.30 – 0.55%
			K	1.80 – 4.00%
			Mg	0.13 – 0.60%
			Ca	0.25 – 1.30%
			Fe	35 – 300 ppm
			Mn	25 – 200 ppm
			B	6 – 40 ppm
			Cu	6 – 40 ppm
			Zn	20 – 70 ppm
			Na	Excessive > 0.8%
Ornamentals				
Abies procera	Noble Fir	Current season's needles.	N	1.4 – 2.0%
			P	0.13 – 0.17%
			K	0.73 – 0.93%
			Ca	0.34 – 0.41%
			S	0.09 – 0.11%
			B	26 – 37 ppm
			Zn	37 – 38 ppm
			Na	Excessive > 0.10%

Latin Name	Common Name	Plant Part for Sampling	Nutrient Sufficiency	Range Low to High
Acer rubrum	Red or Scarlet Maple	Most recent fully matured leaves from new growth.	N	1.9 – 2.4%
			P	0.15 – 0.2%
			K	0.5 – 1.0%
			Ca	0.35 – 2.0%
			S	0.14 – 0.2%
			B	18 – 60 ppm
			Zn	20 – 70 ppm
			Na	Excessive > 0.25%
Agapanthus africanus	Lily of the Nile	Most recent fully matured leaves.	N	2.0 – 3.0%
			P	0.15 – 0.3%
			K	3.0 – 4.0%
			Ca	1.5 – 2.0%
			S	0.14 – 0.3%
			B	20 – 50 ppm
			Zn	25 – 50 ppm
			Na	Excessive > 0.25%
Antirrhinum majus	Snapdragon	Most recent fully matured leaves from new growth.	N	3.5 – 5.0%
			P	0.3 – 0.5%
			K	2.0 – 3.0%
			Ca	1.0 – 2.0%
			S	0.15 – 0.3%
			B	30 – 50 ppm
			Zn	30 – 60 ppm
			Na	Excessive > 0.35%
Araucaria heterophylla	Norfolk Island Pine	Most recent fully matured needles.	N	1.2 – 3.5%
			P	0.15 – 0.45%
			K	1.25 – 3.5%
			Ca	0.5 – 2.3%
			S	0.10 – 0.35%
			B	10 – 65 ppm
			Zn	12 – 200 ppm

(*continued*)

TABLE 11-3 (Continued)

Latin Name	**Common Name**	**Plant Part for Sampling**	**Nutrient Sufficiency**	**Range Low to High**
Arbutus unedo	Strawberry Tree	Most recent fully matured leaves from new growth.	N	1.1 – 2.0%
			P	0.1 – 0.3%
			K	0.75 – 2.0%
			Ca	1.5 – 2.0%
			S	0.1 – 0.2%
			B	30 – 50 ppm
			Zn	30 – 70 ppm
			Na	Excessive > 0.25%
Berberis thunbergii	Japanese Barberry	Most recent fully matured leaves from new growth.	N	2.0 – 4.0%
			P	0.15 – 0.3%
			K	1.0 – 1.5%
			Ca	0.8 – 1.3%
			S	0.15 – 0.3%
			B	30 – 80 ppm
			Zn	30 – 60 ppm
			Na	Excessive > 0.25%
Bougainvillea sp.		Most recent fully matured leaves from new growth.	N	2.4 – 4.5%
			P	0.25 – 0.75%
			K	3.0 – 5.5%
			Ca	1.0 – 2.0%
			S	0.2 – 0.45%
			B	25 – 75 ppm
			Zn	20 – 200 ppm
			Na	Excessive > 0.5%
Buxus microphylla	Japanese Boxwood	Most recent fully matured leaves from new growth.	N	2.3 – 3.5%
			P	0.13 – 0.4%
			K	0.9 – 2.0%
			Ca	1.0 – 2.0%
			S	0.18 – 0.3%
			B	25 – 50 ppm
			Zn	15 – 30 ppm
			Na	Excessive > 0.25%

Latin Name	Common Name	Plant Part for Sampling	Nutrient Sufficiency	Range Low to High
Chamaedorea erumpens	Bamboo Palm	Most recent fully matured leaves from new growth.	N	2.0 – 4.5%
			P	0.11 – 0.75%
			K	1.3 – 4.0%
			Ca	1.0 – 3.3%
			S	0.15 – 1.0%
			B	18 – 60 ppm
			Zn	18 – 200 ppm
			Na	Excessive > 0.25%
Chrysanthemum sp.	Chrysanthemum	Fourth leaf from the growing tip.	N	4.0 – 6.0%
			P	0.25 – 1.0%
			K	4.0 – 6.0%
			Ca	1.0 – 2.0%
			S	0.25 – 0.75%
			B	25 – 75 ppm
			Zn	20 – 250 ppm
			Na	Excessive > 0.5%
Cinnamomum camphora	Camphor Tree	Most recent fully matured leaves from new growth.	N	1.4 – 2.0%
			P	0.15 – 0.3%
			K	1.0 – 2.0%
			Ca	0.8 – 2.0%
			S	0.15 – 0.3%
			B	30 – 80 ppm
			Zn	15 – 60 ppm
			Na	Excessive > 0.25%
Cornus florida	Eastern or Flowering Dogwood	Most recent fully matured leaves from new growth.	N	1.6 – 1.9%
			P	0.15 – 0.55%
			K	0.7 – 1.4%
			Ca	1.6 – 3.6%
			S	0.3 – 0.5%
			B	25 – 70 ppm
			Zn	10 – 35 ppm
			Na	Excessive > 0.25%

(*continued*)

TABLE 11-3 (Continued)

Latin Name	**Common Name**	**Plant Part for Sampling**	**Nutrient Sufficiency**	**Range Low to High**
Cymbidium sp.	Terrestrial Orchids	Most recent fully matured leaves from new growth.	N	1.5 – 2.5%
			P	0.13 – 0.75%
			K	2.0 – 3.5%
			Ca	0.5 – 2.0%
			S	0.15 – 0.75%
			B	25 – 75 ppm
			Zn	25 – 200 ppm
Diospyros kaki	Persimmon	Most recently matured leaves from non-fruiting shoots.	N	1.5 – 2.8%
			P	0.1 – 0.2%
			K	2.0 – 3.8%
			Ca	1.0 – 3.0%
Eucalyptus sp.	Eucalyptus	Most recent fully matured leaves from new growth.	N	1.2 – 3.0%
			P	0.14 – 0.3%
			K	0.7 – 2.0%
			Ca	0.8 – 2.3%
			S	0.15 – 0.3%
			B	30 – 50 ppm
			Zn	16 – 40 ppm
Euonymus alatus	Winged Euonymus	Most recent fully matured leaves from new growth.	N	2.3 – 3.0%
			P	0.20 – 0.45%
			K	1.3 – 2.0%
			Ca	1.0 – 2.0%
			S	0.24 – 0.33%
			B	28 – 40 ppm
			Zn	12 – 34 ppm
			Na	Excessive > 0.25%
Euphorbia pulcherrima	Poinsettia	Most recent fully matured leaves from new growth.	N	3.5 – 6.0%
			P	0.3 – 0.6%
			K	1.5 – 3.5%
			Ca	0.7 – 2.0%
			S	0.1 – 0.3%

Latin Name	Common Name	Plant Part for Sampling	Nutrient Sufficiency	Range Low to High
Euphorbia pulcherrima (continued)			B	30 – 100 ppm
			Zn	25 – 100 ppm
			Na	Excessive > 0.5%
Ficus benjamina	Weeping Chinese Banyan	Most recent fully matured leaves from new growth.	N	1.8 – 3.5%
			P	0.1 – 0.4%
			K	1.0 – 3.0%
			Ca	0.8 – 1.0%
			S	0.14 – 0.4%
			B	20 – 50 ppm
			Zn	15 – 200 ppm
			Na	Excessive > 0.25%
Ficus elastica	Rubber Plant	Most recent fully matured leaves from new growth.	N	1. 3 – 2.5%
			P	0.1 – 0.5%
			K	0.6 – 2.1%
			Ca	0.3 – 1.2%
			S	0.15 – 0.5%
			B	20 – 50 ppm
			Zn	15 – 200 ppm
			Na	Excessive > 0.25%
Fraxinus velutina "Modesto"	Modesto Ash	Most recent fully matured leaves from new growth.	N	2.0 – 2.7%
			P	0.15 – 0.5%
			K	1.0 – 2.0%
			Ca	1.2 – 2.0%
			S	0.14 – 0.3%
			B	30 – 60 ppm
			Zn	15 – 50 ppm
			Na	Excessive > 0.25%
Gardenia jasminoides	Gardenia	Most recent fully matured leaves from new growth.	N	1.5 – 3.0%
			P	0.16 – 0.40%
			K	1.0 – 3.0%
			Ca	0.5 – 1.3%
			S	0.2 – 0.4%

(continued)

TABLE 11-3 (Continued)

Latin Name	**Common Name**	**Plant Part for Sampling**	**Nutrient Sufficiency**	**Range Low to High**
Gardenia jasminoides (continued)			B	25 – 70 ppm
			Zn	20 – 150 ppm
Geranium sp.	Geranium or Cranesbill	Most recent fully matured leaves from new growth.	N	3.5 – 4.8%
			P	0.4 – 0.75%
			K	2.5 – 4.3%
			Ca	0.8 – 1.2%
			S	0.25 – 0.75%
			B	30 – 100 ppm
			Zn	18 – 200 ppm
			Na	Excessive > 0.5%
Gleditsia triacanthos	Thornless Honeylocust	Most recent fully matured leaves from new growth.	N	2.4 – 3.5%
			P	0.14 – 0.3%
			K	1.1 – 2.6%
			Ca	0.67 – 4.0%
			S	0.2 – 0.3%
			B	16 – 40 ppm
			Zn	17 – 50 ppm
			Na	Excessive > 0.25%
Hibiscus rosa-sinensis	Tropical Hibiscus	Most recent fully matured leaves from new growth.	N	2.5 – 3.0%
			P	0.25 – 1.00%
			K	1.5 – 4.5%
			Ca	1.0 – 3.0%
			S	0.2 – 0.5%
			B	25 – 100 ppm
			Zn	20 – 200 ppm
Hydrangea macrophylla	Big Leaf or Garden Hydrangea	Most recent fully matured leaves from new growth.	N	2.9 – 5.5%
			P	0.25 – 0.75%
			K	2.2 – 5.0%
			Ca	0.6 – 2.8%
			S	0.2 – 0.75%
			B	25 – 75 ppm
			Zn	30 – 200 ppm
			Na	Excessive > 0.25%

Latin Name	Common Name	Plant Part for Sampling	Nutrient Sufficiency	Range Low to High
Ilex cornuta	Chinese Holly	Most recent fully matured leaves from new growth.	N	1.8 – 2.8%
			P	0.15 – 0.3%
			K	1.4 – 2.8%
			Ca	0.7 – 1.5%
			S	0.15 – 0.3%
			B	35 – 50 ppm
			Zn	20 – 200 ppm
			Na	Excessive > 0.25%
Lagerstroemia indica	Crape Myrtle	Most recent fully matured leaves from new growth.	N	2.0 – 2.5%
			P	0.14 – 0.3%
			K	0.8 – 2.0%
			Ca	0.65 – 3.0%
			S	0.18 – 0.5%
			B	20 – 90 ppm
			Zn	14 – 60 ppm
			Na	Excessive > 0.25%
Laurus nobilis	Sweet Bay	Most recent fully matured leaves from new growth.	N	2.4 – 3.0%
			P	0.16 – 0.3%
			K	1.0 – 1.6%
			Ca	1.1 – 2.0%
			S	0.1 – 0.2%
			B	25 – 40 ppm
			Zn	30 – 70 ppm
			Na	Excessive > 0.25%
Ligustrum japonicum	Japanese Privet	Most recent fully matured leaves from new growth.	N	2.2 – 3.0%
			P	0.2 – 0.5%
			K	1.6 – 2.5%
			Ca	0.7 – 1.5%
			S	0.2 – 0.4%
			B	20 – 60 ppm
			Zn	20 – 200 ppm
			Na	Excessive > 0.5%

(*continued*)

TABLE 11-3 (Continued)

Latin Name	**Common Name**	**Plant Part for Sampling**	**Nutrient Sufficiency**	**Range Low to High**
Limonium perezii	Sea Lavender or Statice	Most recent fully matured leaves from new growth.	N	2.5 – 3.5%
			P	0.3 – 0.5%
			K	2.2 – 4.0%
			Ca	0.4 – 1.0%
			S	0.24 – 0.5%
			B	20 – 50 ppm
			Zn	24 – 90 ppm
			Na	Excessive > 0.25%
Liquidambar styraciflua	American Sweet Gum	Most recent fully matured leaves from new growth.	N	1.5 – 2.4%
			P	0.14 – 0.35%
			K	0.5 – 1.5%
			Ca	0.6 – 2.0%
			S	0.10 – 0.18%
			B	10 – 50 ppm
			Zn	25 – 40 ppm
			Na	Excessive > 0.25%
Magnolia grandiflora	Southern Magnolia or Bull Bay	Most recent fully matured leaves from new growth.	N	1.6 – 2.2%
			P	0.1 – 0.3%
			K	0.65 – 1.3%
			Ca	0.35 – 0.8%
			S	0.12 – 0.2%
			B	12 – 20 ppm
			Zn	9 – 30 ppm
			Na	Excessive > 0.1%
Morus alba "Fruitless"	Fruitless Mulberry	Most recent fully matured leaves from new growth.	N	2.0 – 4.0%
			P	0.15 – 0.6%
			K	1.3 – 2.0%
			Ca	2.0 – 4.0%
			S	0.15 – 0.3%
			B	30 – 80 ppm
			Zn	20 – 50 ppm
			Na	Excessive > 0.25%

Latin Name	Common Name	Plant Part for Sampling	Nutrient Sufficiency	Range Low to High
Nerium oleander	Oleander	Most recent fully matured leaves from new growth.	N	1.6 – 3.0%
			P	0.2 – 0.3%
			K	2.0 – 3.0%
			Ca	0.7 – 2.0%
			S	0.2 – 0.3%
			B	30 – 90 ppm
			Zn	25 – 70 ppm
Olea europaea	Olive	Basal to mid-shoot leaves.	N	1.5 – 2.0%
			P	0.1 – 0.3%
			K	0.8 – 1.2%
			Ca	1.0 – 1.6%
			B	19 – 150 ppm
			Na	Excessive > 0.20%
Phalaenopsis sp.	Moth Orchid	Most recent fully matured leaves from new growth.	N	2.0 – 3.5%
			P	0.2 – 0.7%
			K	4.0 – 6.0%
			Ca	1.5 – 2.5%
			S	0.2 – 0.7%
			B	25 – 75 ppm
			Zn	20 – 200 ppm
Phoenix canariensis	Canary Island Date Palm	Middle leaflets from most recent fully matured leaves.	N	1.7 – 2.5%
			P	0.15 – 0.3%
			K	1.2 – 1.9%
			Ca	0.6 – 1.5%
			S	0.16 – 0.4%
			B	15 – 60 ppm
			Zn	20 – 200 ppm
			Na	Excessive > 0.25%

(continued)

TABLE 11-3 (Continued)

Latin Name	Common Name	Plant Part for Sampling	Nutrient Sufficiency	Range Low to High
Phoenix dactylifera	Date Palm	Middle leaflets from most recent fully matured leaves.	N	1.7 – 2.5%
			P	0.15 – 0.3%
			K	1. 2 – 1.9%
			Ca	0.6 – 1.5%
			B	15 – 60 ppm
			Na	Excessive > 0.25%
Photinia fraseri	Fraser or Red Top Photinia	Most recent fully matured leaves from new growth.	N	1.2 – 1.8%
			P	0.17 – 0.35%
			K	1.3 – 1.4%
			Ca	1.3 – 2.0%
			S	0.09 – 0.2%
			B	30 – 50 ppm
			Zn	20 – 50 ppm
			Na	Excessive > 0.5%
Pistacia chinensis	Chinese Pistache	Most recent fully matured leaves from new growth.	N	2.1 – 2.8%
			P	0.16 – 0.25%
			K	1.1 – 1.6%
			Ca	0.7 – 1.4%
			S	0.13 – 0.16%
			B	20 – 65 ppm
			Zn	12 – 30 ppm
			Na	Excessive > 0.25%
Pittosporum tobira	Japanese Pittosporum	Most recent fully matured leaves from new growth.	N	1.4 – 3.5%
			P	0.25 – 1.0%
			K	1.5 – 3.5%
			Ca	0.75 – 2.5%
			S	0.2 – 0.4%
			B	20 – 75 ppm
			Zn	20 – 200 ppm
			Na	Excessive > 0.5%

Latin Name	Common Name	Plant Part for Sampling	Nutrient Sufficiency	Range Low to High
Platanus sp.	London Plane Tree or Sycamore	Most recent fully matured leaves from new growth.	N	2.0 – 2.7%
			P	0.13 – 0.25%
			K	0.8 – 1.3%
			Ca	0.5 – 2.3%
			S	0.24 – 0.3%
			B	14 – 86 ppm
			Zn	17 – 42 ppm
			Na	Excessive > 0.25%
Populus sp.	Poplar, Cottonwood, Quaking Aspen	Most recent fully matured leaves from new growth.	N	1.3 – 2.9%
			P	0.13 – 0.4%
			K	1.0 – 2.4%
			Ca	0.6 – 2.3%
			S	0.2 – 0.4%
			B	30 – 80 ppm
			Zn	30 – 200 ppm
			Na	Excessive > 0.25%
Pseudotsuga menziesii	Douglas Fir	Current season's needles from top half of the tree.	N	1.4 – 2.0%
			P	0.15 – 0.35%
			K	0.8 – 1.1%
			Ca	0.25 – 0.5%
			S	0.1 – 0.25%
			B	9 – 40 ppm
			Zn	16 – 45 ppm
			Na	Excessive > 0.1%
Pyracantha sp.	Firethorn	Most recent fully matured leaves from new growth.	N	1.6 – 2.5%
			P	0.13 – 0.3%
			K	0.9 – 1.4%
			Ca	0.9 – 3.0%
			S	0.15 – 0.2%
			B	35 – 70 ppm
			Zn	30 – 70 ppm
			Na	Excessive > 0.25%

(continued)

TABLE 11-3 (Continued)

Latin Name	Common Name	Plant Part for Sampling	Nutrient Sufficiency	Range Low to High
Quercus agrifolia	Coast Live Oak	Most recent fully matured leaves from new growth.	N	1.4 – 2.8%
			P	0.13 – 0.20%
			K	0.65 – 1.4%
			Ca	1.4 – 2.6%
			S	0.13 – 0.20%
			B	15 – 30 ppm
			Zn	15 – 70 ppm
			Na	Excessive > 0.25%
Rhapiolepis indica	India Hawthorn	Most recent fully matured leaves from new growth.	N	1.6 – 2.3%
			P	0.12 – 0.2%
			K	1.4 – 1.5%
			Ca	2.3 – 2.5%
			S	0.09 – 0.2%
			B	30 – 40 ppm
			Zn	60 – 110 ppm
			Na	Excessive > 0.5%
Rhododendron sp.	Azalea	Most recent fully matured leaves from new growth.	N	2.0 – 2.5%
			P	0.3 – 0.5%
			K	0.5 – 1.5%
			Ca	0.7 – 1.5%
			S	0.2 – 0.4
			B	25 – 50 ppm
			Zn	20 – 250 ppm
			Na	Excessive > 0.15%
Rhododendron macrophyllum	Coast or Western Rhododen-dron	Most recent fully matured leaves from new growth.	N	1.3 – 2.1%
			P	0.1 – 0.2%
			K	0.5 – 1.4%
			Ca	0.7 – 1.4%
			S	0.1 – 0.2%
			B	25 – 50 ppm
			Zn	20 – 60 ppm
			Na	Excessive > 0.2%

Latin Name	Common Name	Plant Part for Sampling	Nutrient Sufficiency	Range Low to High
Rosa sp.	Rose	Most recent fully matured five-leaflet leaves.	N	3.0 – 5.0%
			P	0.25 – 0.5%
			K	1.5 – 3.0%
			Ca	1.0 – 2.0%
			S	0.25 – 0.75%
			B	30 – 60 ppm
			Zn	18 – 100 ppm
			Na	Excessive > 0.5%
Salvia splendens	Scarlet Sage	Most recent fully matured leaves from new growth.	N	2.4 – 5.0%
			P	0.3 – 1.0%
			K	2.0 – 5.0%
			Ca	1.0 – 2.5%
			S	0.3 – 0.7%
			B	25 – 70 ppm
			Zn	25 – 80 ppm
			Na	Excessive > 0.25%
Schefflera actinophylla	Queensland Umbrella Tree	Most recent fully matured leaves from new growth.	N	2.0 – 4.5%
			P	0.15 – 0.8%
			K	1.8 – 5.0%
			Ca	1.0 – 2.5%
			S	0.15 – 1.2%
			B	15 – 100 ppm
			Zn	15 – 200 ppm
Spathiphyllum x cultivars	spathi-phyllum or Peace Lilly	Most recent fully matured leaves from new growth.	N	3.3 – 4.5%
			P	0.2 – 0.5%
			K	2.3 – 4.0%
			Ca	1.0 – 2.0%
			S	0.2 – 0.5%
			B	25 – 70 ppm
			Zn	25 – 200 ppm

(*continued*)

TABLE 11-3 (Continued)

Latin Name	Common Name	Plant Part for Sampling	Nutrient Sufficiency	Range Low to High
Syringa vulgaris	Common Lilac	Most recent fully matured leaves from new growth.	N	1.8 – 2.5%
			P	0.25 – 0.4%
			K	1.0 – 2.0%
			Ca	0.6 – 1.2%
			S	0.2 – 0.3%
			B	18 – 40 ppm
			Zn	25 – 75 ppm
			Na	Excessive > 0.25%
Tagetes patula	Marigold	Most recent fully matured leaves from new growth.	N	3.0 – 4.0%
			P	0.3 – 0.7%
			K	2.0 – 3.0%
			Ca	2.0 – 2.8%
			S	0.3 – 0.7%
			B	34 – 60 ppm
			Zn	75 – 100 ppm
			Na	Excessive > 0.25%
Trachelos-permum jasminoides	Star Jasmine	Most recent fully matured leaves from new growth.	N	1.8 – 3.0%
			P	0.18 – 0.5%
			K	1.25 – 2.5%
			Ca	0.75 – 1.5%
			S	0.15 – 0.4%
			B	25 – 75 ppm
			Zn	20 – 200 ppm
			Na	Excessive > 0.5%
Viburnum sp.	Viburnum	Most recent fully matured leaves from new growth.	N	1.4 – 2.5%
			P	0.14 – 0.4%
			K	1.0 – 2.3%
			Ca	0.9 – 2.5%
			S	0.14 – 0.3%
			B	30 – 80 ppm
			Zn	25 – 160 ppm
			Na	Excessive > 0.25%

Latin Name	Common Name	Plant Part for Sampling	Nutrient Sufficiency	Range Low to High
Vinca major & Vinca minor	Periwinkle	Most recent fully matured leaves from new growth.	N	1.6 – 4.0%
			P	0.15 – 0.5%
			K	1.3 – 4.0%
			Ca	0.6 – 1.3%
			S	0.15 – 0.5%
			B	19 – 80 ppm
			Zn	15 – 80 ppm
			Na	Excessive > 0.25%

leaves with distilled water. Washing plant tissue for longer than 10 seconds will leach soluble nutrients such as potassium. Allow the sample to air-dry before packaging. If you package and ship wet leaves or clippings, breakdown of the tissue may occur during shipment. To be safe, do not place leaf tissue in plastic bags. Plant tissue should be placed in paper sacks or laboratory provided paper bags. Ship samples the day of sampling when possible. Where same day shipment is not practical, refrigeration or drying overnight in a warm (200°F) oven can extend the integrity of leaf samples.

The values obtained from plant analysis can be compared to standards only if the sample has been properly collected and if norms are available for the species. Growers should keep accurate records of results and develop their own historical database. As with all proper sampling procedures, be consistent from one sampling period to the next.

Plant analysis is helpful when attempting to diagnose production or quality problems in the nursery or greenhouse. If you wish to compare numbers within a historical database, it is important that leaf samples be taken at the same stage of growth and from the same plant part and location on the plants.

Since a laboratory will need the equivalent of 1 to 2 cups (a large handful) of leaves for analysis, samples from several plants are necessary to provide enough leaves for testing. A number of representative plants should be sampled within the target sampling area.

Where observable problems exist, samples can be collected from areas where the plants are growing normally and from areas where the plants are exhibiting reduced production or poor quality. In this case, two samples would be submitted for testing—one from the healthy plants and one from poorly performing plants. Laboratory results can then be compared for differences.

Unless a specific cause is suspected, a broad range of elements may need to be evaluated. The concentration of nutrients from the laboratory report is compared against norms and used to contrast the normal and affected areas. This is similar to the approach that utilizes soil tests to diagnose production or quality problems.

For best results soil, media, and plant tissue testing should be used in combination to obtain the most diagnostic information for identifying the problem. Irrigation water quality testing and interpretation is also important (refer to Chapter 5). Diagnosing production problems can be complex, and expert advice should be sought for specific problems. Nematodes, diseases, and contaminants, as well as soil water problems, may also be an issue and affect laboratory results.

Plant Tissue Analysis

Plant tissue analysis can be used to identify inadequate, marginal, normal, and excessive concentrations of many nutrients. Plant samples commonly may be analyzed for: total nitrogen, nitrate-nitrogen, phosphorus, potassium, sulfur, sodium, chloride, calcium, magnesium, and micronutrients such as zinc, manganese, iron, copper, boron, and molybdenum. Not all samples need to be analyzed for all these nutrients. Laboratories offer packages that may be less or more comprehensive. Detailed methods of analysis for these tests are available from written sources or by consulting the laboratory.

Interpretation of Analytical Results

Plant analysis results represent the actual concentrations of nutrients found in the plant sample. Plant growth is limited at low nutrient concentrations (see Figure 11-3, page 270). The *plant analysis critical level* occurs at a point where further increases in plant growth cease even though the tissue nutrient concentration is increasing. Plant nutrients at concentrations equal to or greater than the plant analysis critical level will generally not respond positively to an application of that nutrient, while plants with lesser concentrations possibly will. It may be possible to diagnose only one element at a time. For example, in a sample with deficient nitrogen, other nutrients such as phosphorus may accumulate in what would be considered high concentrations. Correction of the most limiting factor may be necessary before the next limiting factor will be detected.

The condition of *hidden hunger* is a low level of plant nutrients within the leaves without observable nutrient deficiency symptoms. *Luxury consumption,* on the other hand, is a condition where a plant takes up fertilizer nutrients above and beyond adequate nutrient levels. Laboratory analysis of plant tissue can identify these conditions that are not visible to the naked eye. Extremely high concentrations of some nutrients may reduce growth due to toxicities, antagonism, or nutrient imbalances.

The concentrations of nutrients found in the plant sample can be compared to published standards and guidelines when they are available. Research on turf and ornamental species to identify these norms has not been as actively pursued as with crop species. Interpretation guidelines may need to be modified and adjusted to allow for differences caused by differing cultural practices, environmental factors, diseases, water stress, and other existing factors.

The nutritional status of trees and shrubs in a wholesale nursery may be evaluated from samples collected at a single specific growth stage or plant part. For plants grown as annuals, such as poinsettia or Easter lilies, it is better to collect samples during the early growing season so trends in the nutrient concentration can be assessed and, if necessary, corrected as soon as possible. The grower should keep historical data to help evaluate plant growth. To make historical data pertinent, sampling procedures and laboratory tests must be consistent each time.

Generalized guidelines for the interpretation of tissue analysis for some ornamental species are given in Table 11-3 (page 278). Nutrient levels are given in relation to the probability of a response from the application of fertilizers supplying a particular nutrient. Tissue analysis values at or below the lower limit of the sufficiency range indicate a high probability of response. Values at or above the upper limit of the sufficiency range indicate a low probability of response. A response may or may not be obtained when tissue test levels are within the sufficiency range. That would depend on production levels and other factors. Concentrations well in excess of the sufficiency range may result from recent foliar sprays or dust. This is very common with micronutrients applied as foliar sprays. Plant tissue that is contaminated with dust or fertilizer should be sent to the lab as soon after sampling as possible since the washing procedure is not effective on dried samples. It may not be possible to wash surface residue of nutrient sprays from leaves. It is better to sample unsprayed leaves.

A tissue analysis, which shows abnormally high or low levels of a nutrient, may not indicate why this has occurred. This is where a soil or media analysis may be helpful in answering why the problem exists. Tissue and soil testing complement one another.

SUPPLEMENTARY READING

Braun, J.R. ed. 1987. *Soil Testing: Sampling Correlation, Calibration, and Interpretation.* Soil Science Society of America, Madison, WI.

Gavlak, R.G., R.D. Horneck, and R. O. Miller. 1994. *Plant, Soil and Water Reference Methods for the Western Region.* Western Regional Extension Publication.

Griffith, L.P. 2007. *Tropical Foliage Plants, Second Edition.* Ball Publishing, Batavia, IL.

Jones, J.B., Jr., Wolf, B., and Mills, H.A. 1996. *Plant Analysis Handbook. A practical sampling, preparation, analysis, and interpretation guide.* MicroMacro Publishing, Inc., Athens, GA.

Kalra, Y. 1998. *Handbook of Reference Methods for Plant Analysis.* CRC Press, LLC, Boca Raton, FL.

Kopittke, P.M., and N.W. Mendzies. 2007. A review of the use of the basic cation saturation ratio and the "ideal" soil. Soil Sci. Soc. Am. J. 71 259-265.

Page, A.L., ed. 1982. *Methods of Soil Analysis, Part II.* American Society of Agronomy, Madison, WI.

Perry, E., and G. Hickman. 2001. *Fertilizing Landscape Trees.* University of California, Publication 8045.

Soil Testing and Plant Analysis Council. 1999. *Soil and Plant Analysis Laboratory Registry for the U.S. and Canada, Second Edition.* CRC Press, LLC, Boca Raton, FL.

Westerman, R.L. ed. 1990. *Soil Testing and Plant Analysis.* Soil Science Society of America, Madison, WI.

CHAPTER 12

Nutrient Guidelines

Careful nutrient management is important for responsible fertilization of ornamental plantings, sports fields, parks, or any other managed landscape. The goal of ensuring good growth and attractive appearance must be balanced with minimizing costs and environmental impacts. Fertilization starts with basic guidelines and then adapts the *Four Rights (4Rs)* of applying the right product, at the right rate, at the right time, and in the right place for the existing conditions. Many factors must be considered when choosing and applying fertilizers to supply required nutrients. The following guidelines provide starting points for responsible nutrient management of turf and ornamental plantings.

TURFGRASS NUTRIENT MANAGEMENT

Turfgrass is found in diverse areas such as residential and commercial lawns, parks, cemeteries, sports fields, schoolyards, roadsides, and golf courses, to name a few examples. Segmenting turf into amenity and functional elements helps to clarify the wide range of performance demands supported by distinct fertilizer programs. *Amenity turf* is grass planted and maintained for primarily aesthetic reasons. Amenity turf includes lawns, parks, and formal gardens. Amenity turf is generally successful on unamended soils. *Functional turf* is grass planted and maintained for play and high traffic areas, such as sports fields. Functional turf can be segregated into four levels of quality to meet performance expectations: premium, choice, standard, and play.

Fertilizer application varies with expected quality level and field construction. Functional turf may be established on unamended soils, but performance is higher when the root zone is modified for compaction resistance and improved drainage. The overall complexity of turf culture is only apparent when considering the different uses, climate, wide variation in turfgrass species, and the difference between native (field) soils and modified root zone media.

Effect of Soils on Turf Performance

Turf may be established on soils ranging from well-drained, coarse-textured loams and sandy loams to fine-textured clays with poor structure and drainage. Unamended soils are generally subject to compaction from foot traffic. Turf which is growing on compacted soil may not respond to fertilization. Alleviation of compacted soils is a common cultural practice along with irrigation and fertilization.

Amendments often improve the water-holding capacity, decrease bulk density, improve resistance to compaction, improve root penetration, and increase cation exchange capacity of soils under turf. Organic amendments such as peat, weathered sawdust, and green waste compost are used to varying degrees. Mineral amendments such as calcined clay, ceramic clay, and sand, when used in sufficient volume, can increase porosity of some fine textured soils.

Root zone mixtures are familiar to most turf managers. Golf putting greens and sports fields have specific functions that benefit from the development and use of root zone mixtures. Pure sand with a narrow range particle size distribution has wide acceptance in golf and sports turf management. Mixtures of soil, sand, and various mineral and organic materials are also quite common. These mixtures are primarily designed to resist compaction and provide drainage, while retaining a consistent water holding capacity. Refer to Chapter 6 for more detailed information on soil amendments.

Turf Fertilization Programs

Nitrogen

Turfgrass requires nitrogen (N) in the largest amount of any of the essential nutrients. A suggested range for N in turfgrass tissue is 2.0 to 6.0 percent depending on the species (see Table 11-3, page 278).

From a practical perspective, desired turf color and rate of growth generally influences decisions on rates and timings of N fertilizer applications. Increasing N application, within limits, generally increases the above ground biomass. This biomass creates a cushion and mat that adds resilience to the surface and increases traffic tolerance. However, the root-to-shoot ratio decreases with increasing N, which increases the cushion but reduces the recuperative potential of the grass. Plants entering the summer stress period with reduced root development may have increased succulence and are more susceptible to disease.

Nitrogen application not only increases shoot biomass, it also promotes chlorophyll formation, leading to a greener appearance of the turf. Excessive buildup of biomass, usually as thatch, can reduce water infiltration and air exchange. Nitrogen leaching losses from well-managed turf are generally quite low. The dense root system increases nutrient recovery and minimizes N losses. However, when applied in excess, the economic losses and environmental harm caused by inefficient N utilization can be severe.

Nitrogen fertilization programs differ between warm and cool season grass species. Warm season grasses are those species that essentially stop growing when root zone temperature at 2 inches of depth is less than 50°F and also grow slowly when the root zone temperature exceeds 95°F. Cool season grasses stop growing when root zone temperature at 2 inches of depth is below 40°F and above 80°F. Fertilizer should not be applied when root zone temperatures exceed these thresholds. The most common warm and cool season grass species in the Western United States are listed in Table 12-1.

The growth rate of warm season grasses peaks in midsummer. Ideal nitrogen fertilizer application timing occurs before the peak growth period, but not during periods of extreme heat or cold (Table 12-2). These generalized warm and cool season programs should be modified to fit the needs of the root zone media and climate. Additional fertilizer applications may be necessary in those programs with less than 4 pounds N/1,000 sq ft/year, depending on visual and growth requirements.

TABLE 12-1

Warm and Cool Season Grass Species

Warm Season		Cool Season	
Common Name	**Latin Name**	**Common Name**	**Latin Name**
Bermudagrass	*Cynodon dactylon*	Kentucky Bluegrass	*Poa pratensis*
Hybrid Bermudagrass	*Cynodon dactylon x C. transvaalensis*	Perennial Ryegrass	*Lolium perenne*
Japanese Lawngrass	*Zoysia japonica*	Tall Fescue	*Festuca arundinaceae*
St. Augustine Grass	*Stenotaphrum secondatum*	Creeping Bentgrass	*Agrostis palustris*
Buffalograss	*Buchloe dactyloides*	Creeping Red Fescue	*Festuca rubra*
Seashore Paspalum	*Paspalum vaginatum*	Hard Fescue	*Festuca ovina*
		Colonial Bentgrass	*Agrostis tenuis*
		Annual Ryegrass	*Lolium multiflorum*
		Roughstalk Bluegrass	*Poa trivialis*

TABLE 12-2

A Warm Season Turfgrass Nitrogen Fertilization Program

	(lb/N/1,000 sq ft)												
Uses	**J**	**F**	**M**	**A**	**M**	**J**	**J**	**A**	**S**	**O**	**N**	**D**	**Total**
Golf Greens		1		1	1	1	1	1	1	1			8
Commercial Lawns		1		1		1	1	1		1			6
Sports Fields													
Premium		1		1	1		1	1	1	1	1		8
Choice		1		1		1	1	1		1			6
Standard				1		1		1		1			4
Play				1						1			2
Home Lawns				1		1				1			3
Parks/Schools				1						1			2
Highways				1									1

TABLE 12-3

A Cool Season Turfgrass Nitrogen Fertilization Program

Uses	(lb/N/1,000 sq ft)												
	J	F	M	A	M	J	J	A	S	O	N	D	Total
Golf Greens		1	1	1	1				1	1	1	1	8
Commercial Lawns		1		1	1				1			1	5
Sports Fields													
Premium		1	1	1	1	1			1		1	1	8
Choice		1		1	1				1			1	5
Standard				1		1				1			3
Play				1						1			2
Home Lawns		1			1				1				3
Parks/Schools				1						1			2
Highways				1									1

Cool season grasses have two growth rate peaks during the year (Table 12-3). The first peak starts in the spring as the soil warms. The growth rate increases rapidly until the end of June. The mid-summer growth rate drops considerably. A second period of growth increase begins in September. The N fertilizer application schedule is appropriate for regions where cool-season grasses remain active throughout the year (e.g. most of California). In regions that experience a distinct growing season, such as the Intermountain West, Pacific Northwest, and high elevations of Arizona, California, Nevada, and New Mexico, application timing should be adjusted to the periods of active growth.

In general, the suggested N application rates can be reduced between 25 to 35 percent if clippings are returned to the turf through mulching (Table 12-4).

Phosphorus

Phosphorus (P) is taken up by turfgrass in relatively high quantities. Phosphorus in mineral soils is generally insoluble, resists leaching, and remains close to where it is applied. However, phosphorus is known to leach from the root zone in very sandy soils or in peat-based media. Phosphorus becomes increasingly

TABLE 12-4
Average Concentration of Nutrients in Municipal-Collected Grass Clippings

Nutrient	Pounds/Dry Ton	Nutrient	Parts Per Million (dry)
Nitrogen	61	Iron	5,200
Phosphorus (P_2O_5)	18	Manganese	130
Potassium (K_2O)	46	Copper	24
Sulfur	8	Zinc	85
Calcium	17		
Magnesium	7		

Krogmann, U., J.R. Heckman, and L.S. Boyles. 2001. Nitrogen mineralization of grass clippings – A case study in fall cabbage production. *Compost Sci.* 9:230–240.

unavailable to the plant if soil pH is greater than 7.5 or below 5.0. Phosphorus-deficient turf frequently appears stunted and may show a reddish purple color beginning at the leaf tips. Soil-extractable (Olsen) P concentrations of 15 to 30 ppm phosphate-P in the soil and 0.3 to 0.6 percent P in leaf tissues are commonly sufficient. Turf managers should keep soil P in the upper part of this range to avoid limiting plant growth, especially on high traffic areas. On both cool and warm season grasses, P is applied at the rate of 0.5 to 1.0 lb P_2O_5/1,000 sq ft/year. There is scientific evidence that applying P once per year in conjunction with N and K promotes growth. There is a benefit to root development in applying P to new seedlings and under sod in new sod installations.

Potassium

Potassium (K) improves tolerance of turfgrass to wear, drought, heat stress, and disease. Potassium also adds to aesthetic quality. Soil-extractable K concentrations of 100 to 250 ppm K in the soil and 1.0 to 3.5 percent K in leaf tissues are generally sufficient for turf growth. Optimum wear tolerance is achieved at 6 to 8 lb K_2O/1,000 sq ft/year if N is not deficient. Sand-based modified root zone mixtures on golf greens and athletic fields present a special challenge for K availability because of low nutrient holding capacity.

Reduced rates and frequent applications of K in balance with N applications are recommended.

Fertilizing for Turf Performance

Turf density is a function of turf vigor, which is an indication of the ability of roots, rhizomes, and tillers to recover from injury. Maintenance of adequate concentrations of P in the root zone, with balanced concentrations of N and K, allow the plant to respond to injury.

Turf with adequate N and a balance of P and K has maximum stress resistance. When using only high N application rates to build biomass, the susceptibility to drought, heat, cold, and traffic increases. Resistance to these stresses can be improved with balanced applications of any essential element that may be lacking, with special attention to P and K.

Relatively high N rates, common for golf and sports turf, may increase the incidence of some diseases while offering resistance to others. Improving internal drainage with a sand-based root zone mixture helps with disease resistance when high rates of N are applied. Maintaining adequate concentrations of K may reduce the potential for disease injury during hot weather.

An application of N frequently improves turf color. Iron (Fe) may also intensify green color. If N is adequate, an application of a soluble form of Fe fertilizer, such as ferrous sulfate heptahydrate at rates of 0.25 to 0.5 lb Fe/1,000 sq ft/application, will frequently darken the color of the turf within two to three days. Irrigation should immediately follow an application of ferrous sulfate because it can damage the turf if allowed to remain on the foliage. Application of ferrous sulfate on a hot day or at a high rate may damage the turf where applicator tires scuff the turf. Ferrous sulfate may also stain concrete walkways and should be applied with care.

Late season N applied before warm-season grasses go dormant improves fall color retention. Nitrogen applied before the last mowing of cool-season grasses improves spring green-up. A controlled-release form of N, either alone or blended with other nutrients, can reduce the potential for losses of applied N through leaching, volatilization, or run-off. Late-season applications must be made early enough to allow complete uptake of the applied N before turf goes dormant in order to avoid N losses.

Turf Facilities

For facilities management purposes, amenity turf and functional turf are further segmented by the cultural intensity required. The segments are: golf turf, commercial lawns, sports fields, home lawns, memorial parks, parks and schools, and highway roadsides.

The most highly-managed functional turf is the putting green. Consistent growth for high turf density is important for golf greens. Fertilizer is applied regularly to greens throughout the growing season. The most widely used cool season grass for putting greens is creeping bentgrass. High-density bentgrasses can tolerate very low mowing heights and receive much less N to reduce thatch buildup. High-density greens may only receive 5 to 6 lb N/1,000 sq ft/year. Fertilizer applications are usually not made during the mid-summer due to high temperatures.

Hybrid bermudagrass greens have similar fertilizer requirements as bentgrass greens. However, hybrid bermudagrass responds well to summer fertilizer applications. Hybrid bermudagrass greens are often over-seeded with a cool season grass for winter color, usually perennial ryegrass or roughstalk bluegrass, in late October or early November. At least 2 lb N/1,000 sq ft during the fall and winter is applied for the overseeded species. Overseeded turf then transitions back to bermudagrass in late spring or early summer with the high temperatures.

On sand greens, an equal amount of K should be made for each unit of N applied. On greens with amended native soil, K should be applied at 6 to 8 lb K_2O/1,000 sq ft/year and P applied at least at 1 lb P_2O_5/1,000 sq ft once per year.

"Push up" greens are initially planted in unamended soil. Over time, root zones are modified through aeration, topdressing, or some other process. Fertilizer programs must be customized for each green. Pure sand greens have greater potential to leach soluble N. Eventually, the turfgrass root mass will become sufficient to capture and utilize the added N. These greens typically receive 6 to 8 lb N/1,000 sq ft/year and 1 lb P_2O_5/1,000 sq ft once per year. Nitrogen rates should be reduced to prevent excess top growth in relation to root growth as needed. Potassium requirements are often lower for turf growing on native soils than on sand greens.

Tees and fairways must have "playability" and high traffic tolerance. Fertilizer recommendations for tees and fairways are similar to that for commercial lawns. During the growing season, cool season grasses may require 5 lb N/1,000 sq ft/year, and warm season grasses may require as much as 6 lb N/1,000 sq ft/year. Par three tees and approach areas of fairways are subject to extensive divotting and may respond to increased N. These two high impact areas may also respond to increased levels of K to improve traffic tolerance and encourage growth for injury recovery.

Commercial lawns are generally considered amenity turf. Desirable characteristics of commercial lawns are deep green color, stand density, and uniformity. Cool season grasses are fertilized primarily in the spring and fall, while warm season grasses are fertilized in the spring and summer. For commercial lawns that also have a functional application, such as golf course fairways, 5 lb N/1,000 sq ft/year is applied to cool season grasses during the growing season and warm season grasses may receive as much as 6 lb N/1,000 sq ft/year.

ORNAMENTAL NUTRIENT MANAGEMENT

Unlike turf, it is difficult to give specific fertilizer programs for ornamental plantings due to the great diversity of species and nutrient requirements in the landscape. However, there are several general fertilizer management guidelines that will provide environmentally safe and economically sound fertilizer use while maintaining aesthetically pleasing ornamental plants.

Fertilizer applied to landscape ornamental plantings is very important in enabling plants to develop to their fullest genetic potential. The type and amount of fertilizer applied is directly related to several factors. The age or stage of development of the planting needs to be considered. A new planting where the plant material is still undersized, compared to the ultimate expectations of the planting, will require more fertilizer than a mature planting where maintaining the size would be preferable to stimulating additional vegetative growth. Soil type and texture influence how much and how frequently to apply fertilizer. Plants growing in coarse-textured soil will do better when small amounts of fertilizer are applied frequently. Less frequent applications of fertilizer at higher rates can work well with soil that has a high percentage of silt and clay.

Time of year and soil temperature also affect fertilizer selection. In the fall, applying higher amounts of nitrogen may stimulate new vegetative growth, which is more susceptible to frost or cold damage. Applying additional potassium frequently increases the cold tolerance of plants. Additionally, when nitrogen is applied in the ammonium form when soil temperatures are below 50°F, it will be less available to plants due to sluggish conversion to nitrate in cold soil. The percent of nitrogen in ammonium and nitrate forms in a fertilizer can be found on the label. In the cool season, it may be beneficial to use fertilizer with a higher proportion of nitrate to ensure rapid response.

In many situations, fertilizer applied to turfgrass areas will be adequate for adjacent and surrounding ornamental plantings. Returning grass clippings to the soil can contribute significant quantities of nutrients when they decompose and become recycled (Table 12-4, page 302). Ornamental plantings in a landscaped area comprised of both turf and ornamentals may need less fertilizer than the turf to maintain good plant appearance.

The growing media used to grow plants also affects the fertilizer requirements. *Specimen trees* and shrubs are focal-point, isolated, ornamental perennials that have been grown in unamended field soil and brought into the landscape planting in either large containers or as "ball and burlap" plants. Specimen trees will need much less fertilizer during their establishment period than container plants growing in highly-amended nursery soil mixes. Bedding plant annuals and many perennials often have higher fertilizer requirements because of their rapid growth rate and dominant position in most plantings. It is often appropriate to supplement bedding plants with additional fertilizer during the growing season.

Some ornamental plants not native to the Western states may require acid-forming fertilizers in order for them to thrive in landscape plantings. Included in this group are azaleas, rhododendrons, camellias, gardenias, some ferns, and blueberries. These plants do well in highly-amended organic soil mixes that are low in salinity (EC_e less than 2.5 dS/m) and have a moderately acidic (pH 4.5 to 5.5) reaction.

Applying fertilizer in irrigation water, or fertigation, has been successfully used in nursery and greenhouse environments for many years. Fertigation is easily managed and can be programmed so the plants receive small amounts of fertilizer with irrigation water when desired. A fertigation system for a nursery or greenhouse consists of equipment in the water line to measure water flow, tanks to hold fertilizer stock solutions, and pumps to add the correct amount of fertilizer when needed. In many current systems, sensors are included to monitor the amount of fertilizer present in the line and even make adjustments in response to the measurements.

Many nursery and greenhouse facilities are set up as "closed" systems so no runoff water leaves the property. Instead, runoff is treated with chlorine or ozone to reduce or eliminate water-borne plant pathogens and tested to determine what additional fertilizer should be added before it is recycled back into the irrigation water. If the total salt level of the recycled runoff becomes excessive, it is possible to dilute the recycled water with additional fresh water or in some situations, remove excess salts with special filter systems such as reverse osmosis. Sometimes fertigation systems are also included in the plans for landscape plantings. Examples of typical liquid fertilizer formulations for several ornamental production systems are shown in Table 12-5.

Soil and plant tissue samples are useful in monitoring the success of fertilizer programs. With new plantings, soil samples should be analyzed before the plants are installed. During residential construction, fertile topsoil is often removed and used as fill when soils are leveled. As a result, the native subsoil, which is typically very low in organic matter and fertility, ends up at the surface in the planting areas. The use of heavy construction equipment during development can also compact native soils. A soil test prior to planting will identify the existing conditions and provide a practical plan for transforming this soil into a suitable growing medium for the intended landscape plantings. Typical recommendations may suggest incorporating 1 to 2 inches of suitable organic matter along with appropriate fertilizers and chemical amendments like gypsum, sulfur, or lime to improve the soil to the point where only maintenance fertilizer will be needed. These amendments and fertilizers are typically incorporated to a depth of 6 inches in the planting areas.

TABLE 12-5

Typical Concentrations of Several Liquid Fertilizer Programs[1,2]

Plants Being Grown	Nutrients Needed	Concentration of Nutrients
Plugs and Liners	Nitrate N + K + Micros	50-150 ppm N
		50-100 ppm K
		EC 0.4-1.1 dS/m
Greenhouse Potted Plants	25% NH_4–N, 75% NO_3–N	100-300 ppm N
	K	100-200 ppm K
	EC	0.8-2.1 dS/m
Cut Flowers	25% NH_4–N, 75% NO_3–N	100-300 ppm N
	K	100-200 ppm K
	+K + Micros	EC 0.7-1.4 dS/m
Nursery Stock	70% Nitrate N	75-200 ppm N
	30% Ammonium N	50-125 ppm K
	+K + Micros	EC 0.5-1.3 dS/m
Bedding Plants	75-100% Nitrate N	75-150 ppm N
	0-25% Ammonium N	50-100 ppm K
	+K + Micros	EC 0.5-1.2 dS/m
Landscape Turf & Groundcover	50% Nitrate N	35-75 ppm N
	50% Ammonium N	25 ppm K
	+K + Micros	EC 0.3-0.5 dS/m

[1]Phosphorus may be required in any of these programs depending upon the chemical properties of the growing medium.

[2]Ratio between NO_3–N and NH_4–N depends upon crop, time of year, soil temperatures, and weather conditions. Most research indicates that for most crops NH_4–N should not exceed 25 to 30 percent of total N.

Maintenance fertilizer rates for nitrogen are similar to those used for turf. The application rates for phosphorus and potassium are generally lower for established landscape plantings. A fertilizer with an N–P–K ratio of 3-1-2 is often appropriate for landscape plantings where the soil has been adequately prepared prior to planting.

Fertilizer requirements will vary depending upon how the plants will be used after purchase. For example, in a commercial greenhouse or nursery environment, the goal is typically to produce a high-quality plant of saleable size as rapidly as possible. This usually requires an aggressive fertilizer program. After the same plants are purchased by the consumer and transplanted into larger containers to be used for decoration, it may be desirable to control the growth but still maintain good plant appearance for an extended period of time. This can be accomplished by reducing the rate of fertilizer application. If the plants purchased from the nursery are replanted into field soil with the intent of producing large landscape specimens, a fairly aggressive fertilizer program should be maintained based upon the fertility of the root zone soil.

Nutrient Management for Shrubs and Landscape Trees

The fertilization of shrubs and landscape trees is often neglected. Many people assume that fertilization of adjacent flower beds or turf will also take care of the trees and bushes growing next to them. This is not always correct. The fertilizer amounts, application methods, and timing for ornamental beds and turf are usually targeted to benefit the flowers or the grass, not the larger woody plants.

Evergreen trees, such as conifers and citrus, should be fertilized before new vegetative growth appears, generally in the spring. Deciduous trees and shrubs should be fertilized in early spring, as soon as their buds are starting to swell. This indicates that dormancy is breaking, root growth has resumed, and that nutrients can be taken up again from the soil. Apply fertilizers inside the drip line of the tree or shrub. Surface application is effective where there are no other plants growing underneath the tree or shrub, such as when there is a mulched tree ring around the trunk. In the case of turf or plant beds under the tree canopy, fertilizer should be applied in holes at least 6 inches below the soil surface. Otherwise, the turf or bedding plants will take up the majority of fertilizer before it has a chance to reach the roots of the tree or shrub. This can be accomplished by using a soil probe and dropping the fertilizer into the holes created. Another method is to open a crack in the soil at a slight angle with a spade or shovel, drop some fertilizer into the crack, withdraw the implement, and close the crack.

The annual rate of nitrogen application depends on the area inside the tree's dripline. Depending on tree vigor and desired vegetative growth, apply 2 to 4 lb N/1,000 sq ft/year. To find the size of a tree's canopy-covered area, measure the longest and widest extent of the drip line in feet and multiply the two measurements. Divide the area by 1,000 and multiply this result by the desired nitrogen application rate to obtain the total amount of nitrogen required for that tree.

Established trees and shrubs generally do not require annual fertilization with phosphorus or potassium. Every 3 to 5 years, apply 1 to 3 lb P_2O_5 and K_2O/1,000 sq ft together with the annual nitrogen fertilization. However, trees and bushes with desirable ornamental flowers, such as lilac or magnolia, benefit from the use of complete balanced fertilizers and will flower more profusely if phosphorus is applied at rates equal to or higher than nitrogen.

FERTILIZER BEST MANAGEMENT PRACTICES

Fertilizer Best Management Practices (BMPs) are practical actions that can be implemented to promote the efficient use of nutrient resources and minimize adverse environmental impacts. They are commonly promoted to protect natural resources, such as air, water, and soil. Appropriate BMPs are used to maintain productivity and enhance long-term sustainability of the production area and the environment. At a practical level, plant production systems are managed for multiple objectives. Best Management Practices are selected to closely achieve those objectives. The objectives include a complex mix of economic, ecological, and societal goals.

On a more practical basis, fertilizer BMPs are a subset of plant management BMPs. If a fertilizer practice is to be considered a "best" practice, it must align with profitability, productivity, environmental, and sustainability goals.

The Four Rs

One approach to fertilizer BMPs is to consider the Four Rs (Four Rights) nutrient stewardship concept (Figure 12-1). This concept implies that the right nutrient product is applied at the right rate, right time, and right place. This simple framework can help managers understand how to select the right management practices so that fertilizer can contribute to sustainability (Table 12-6).

In past years, a team of growers, advisers, university staff, and agribusiness professionals would decide what constituted a BMP. However, additional stakeholders are now influencing how nutrient decisions are being made—including people consuming horticultural products or living in the environment that they impact. Multiple objectives must now be considered every time a nutrient decision is made.

The phrase "right product at the right rate, right time, and right place" may imply that there is one correct practice, or

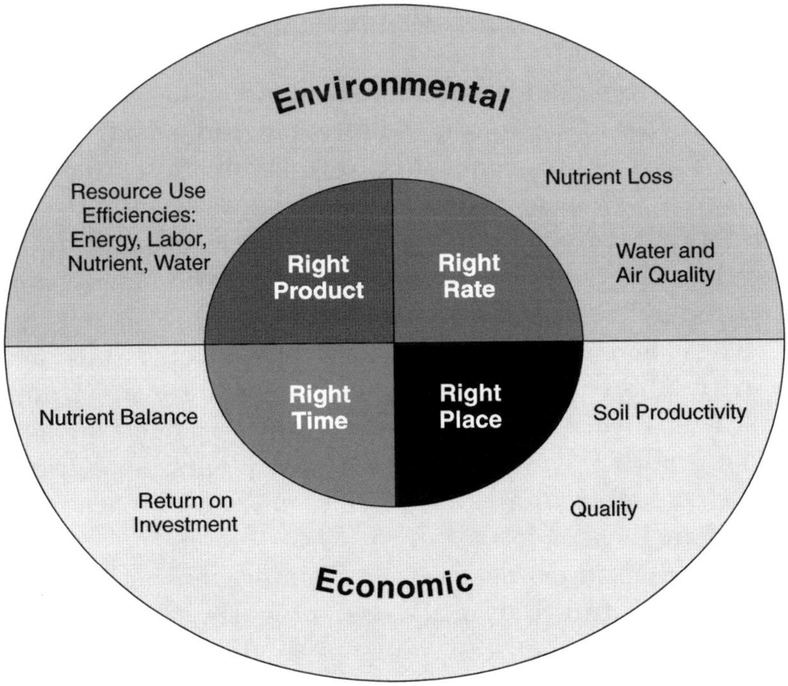

FIGURE 12-1. The Four Rs nutrient stewardship concept for deciding Best Management Practices. *Source:* Adapted from The Fertilizer Institute.

TABLE 12-6
General Principles for Developing Practices to Determine the Four Rs

Right	Principle
Product	Material(s) chosen should provide an adequate supply of all essential nutrients, considering both naturally available sources and characteristics of specific products in plant-available forms.
Rate	Assess the nutrient supply and plant demand, use laboratory analysis when needed.
Time	Assess the dynamics of plant nutrient uptake, the existing nutrient supply, and the logistics of field operations. Consider how to avoid nutrient loss.
Place	Recognize root and soil interactions, account for spatial variability in a landscape or container to best meet plant needs. Consider the advantages and disadvantages of nutrients applied before and after planting in soil and media, foliar applications, and fertilizer in irrigation water.

group of practices, that should be widely adopted. However, these principles need to be locally adapted to individual situations. They provide a guiding checklist to consider: "Am I using every tool available to choose the right material, to predict the proper material application rate, to apply it at the right time, and place it where it is most effective for my plants, soil conditions, weather, and consumer?"

These Four Rs are completely interconnected and should be linked with all the management practices for the entire production system. This involves a complicated interaction of the physics, chemistry, and biology of the nutrition of plants. Local expertise should be consulted to make sure that these BMPs are in accord with the Four Rs principles.

This chapter cannot present a comprehensive list of BMPs for every horticultural situation, but provides some examples of fertilizer practices that can promote environmental, economic, and social goals associated with plant production. Practices that are appropriate in one situation may be entirely inappropriate or even damaging in another location. Local expertise must be used to select the practices most suitable for a given situation.

The following list presents some suggested BMPs for greenhouse and container nursery production.

- Staff training to communicate expectations, maintain up-to-date information, and monitor progress
- Proper storage of fertilizer material and equipment
- Maintenance and calibration of injector equipment
- Limit loss of nutrients from the surface of containers during watering
- Minimize water and nutrient loss during irrigation
- Use plant tissue analyses to determine need for nutrient application
- Choose fertilizer materials and timing based on plant requirement
- Select most appropriate growth media
- Collect, treat, and recycle runoff water as appropriate

The following list presents some suggested BMPs for turf management.

- Utilize proper turf species for the intended purpose
- Minimize sediment loss during the grow-in period
- Apply fertilizer several times a year, instead of a single large application
- Avoid fertilizer application on impervious surfaces, such as parking lots and streets
- Calibrate fertilizer spreaders to apply the correct amount and pattern
- Establish buffers around sensitive areas that should not be fertilized
- Use proper mowing height and manage clippings appropriately
- Use plant tissue analyses to determine need for nutrient application
- Only use nutrients that are needed to sustain healthy turf function

In brief, the Four Rs stewardship nutrient concept involves users selecting the right product-rate-time-place combination from practices validated by research. Use the appropriate performance indicators to meet local goals (Table 12-7). All nutrient users have an obligation to manage these resources as carefully as possible. Implementation of appropriate BMPs can result in greater efficiency, profitability, and sustainability.

TABLE 12-7

Examples of Performance Measures and Indicators for Fertilizer Management Practices[1]

Performance Measure or Indicator	Description
Yield	Number of plants harvested per unit planted area per unit of time.
Quality	Appearance, storage life, sugar, protein, minerals, vitamins, or other attributes that add value to the harvested product.
Nutrient use efficiency	Biomass produced or nutrient taken up per unit of nutrient applied.
Water use efficiency	Units of water applied versus units of water consumed by plants.
Labor use efficiency	Labor productivity, linked to number, and timing of field operations.
Energy use efficiency	Plant production per unit of energy input.
Net profit	Volume and value of plants produced relative to all costs of production. Return on investment = profit in relation to capital investment.
Adoption	Proportion of producers using particular BMPs.
Soil productivity	Soil fertility levels, soil organic matter, and other soil quality indicators.
Yield stability	Resilience of plant yields to variations in weather and pests.
Operation income	Improvements in livelihood.
Working conditions	Quality-of-life issues, worker satisfaction, employee turnover, and worker safety.

(continued)

Performance Measure or Indicator	Description
Water and air quality	Nutrient concentration and loading in groundwater, watersheds, or airsheds.
Ecosystem services	Countryside aesthetics, natural predators and pollinators, outdoor recreation, hunting, fishing, etc.
Biodiversity	Difficult to quantify—can be descriptive.
Soil erosion	Degree of soil coverage by actively growing plants and plant residues, relative to soil losses with wind and water.
Off-field nutrient losses	The combined total of nutrient losses from the management zone, edge of field, bottom of root zone, and top of crop canopy.
Nutrient budget	A total account of nutrient inputs and outputs, at the soil surface or plant production operation.

[1]The relative importance among these and other indicators needs to be determined by stakeholder input.

SUPPLEMENTARY READING

Beard, James B. 2002. *Turf management for golf courses. 2nd ed.* A publication of the United States Golf Association. Ann Arbor Press, Ann Arbor, MI.

Bruulsema, T. et al. 2008. A global framework for best management practices for fertilizer use. International Plant Nutrition Institute. www.ipni.net/4r.

Colt, M., T. Tindel, D. Barney, and B. Tripepi. 1997. Fertilizing landscape trees. Univ. of Idaho Cooperative Extension Publication No. CIS 1068.

Pessarakli, M. ed. 2008. *Turfgrass Management and Physiology.* CRC Press Taylor & Francis Group, Boca Raton, FL.

APPENDIX A

Useful Tables and Conversions

TABLE A-1

Useful Weights and Measures for the Gardener and Horticulturist

Weights

Pounds per Acre	Equivalent Quantity per 100 Square Feet	Equivalent Quantity per 1,000 Square Feet
100	3½ ounce (oz)	2 lb 5 oz
200	7½ oz	4 lb 9 oz
300	11 oz	6 lb 14 oz
400	14 ¾ oz	9 lb 3½ oz
500	1 lb 2½ oz	11 lb 8 oz
600	1 lb 6 oz	13 lb 12 oz
700	1 lb 10 oz	16 lb 1 oz
800	1 lb 13 oz	18 lb 6 oz
900	2 lb 1 oz	20 lb 11 oz
1,000	2 lb 5 oz	22 lb 15 oz
2,000	4 lb 10 oz	45 lb 15 oz

(*continued*)

TABLE A-1 (Continued)

Measures (approximate)

1 level teaspoon (tsp)	⅙ oz
1 level tablespoon (tbsp)	½ oz
1 level cup	8 oz
1 pint (pt)	1 pound (lb)
1 quart (qt)	2 lb
1 gallon	8 lb

TABLE A-2
U.S. Weights

Troy Weight

24 grains (gr)	=	1 pennyweight (pwt or dwt)
20 pennyweight	=	1 troy ounce (fl oz)
12 troy ounces	=	1 troy pound (lb)

Apothecaries' Weight

20 grains (gr)	=	1 scruple (sc)
3 scruples	=	1 dram (dr)
8 fluid drams	=	1 fluid ounce (fl oz)
12 ounces	=	1 pound (lb)

Avoirdupois Weight

27 11/32 grains (gr)	=	1 dram (dr)
16 drams	=	1 ounce (oz)
16 ounces	=	1 pound (lb)
25 pounds	=	1 quarter
4 quarters or 100 pounds (U.S.)	=	1 hundredweight (cwt)
112 pounds (Great Britain)	=	1 hundredweight
20 hundredweight or 2,000 pounds (U.S.)	=	1 ton (t)*

*The U.S. short ton (st) is 2,000 lb, the British long ton (lt) is 2,240 lb, and the metric tonne (T) (1,000 kg) is 2,204.6 lb. The long ton is also used in the United States as a measure of weight, especially by steamship companies and customs.

TABLE A-3

Length and Area (Land Measurements)

Linear Measure	Square Measure
1 inch (in) = 0.0833 foot (ft)	144 sq in = 1 sq ft
7.92 in = 1 link	9 sq ft = 1 sq yd
12 in = 1 ft	30¼ sq yd = 1 sq rd
3 ft = 1 yard (yd)	16 sq rd = 1 sq ch
25 links = 16½ ft	1 sq rd = 272¼ sq ft
25 links = 1 rod (rd)	1 sq ch = 4,356 sq ft
100 links = 1 chain (ch)	10 sq ch = 1 acre
16½ ft = 1 rd	160 sq rd = 1 acre
5½ yd = 1 rd	4,840 sq yd = 1 acre
4 rd = 100 links	43,560 sq ft = 1 acre
66 ft = 1 ch	640 acres = 1 sq mi
80 ch = 1 mile (mi)	1 sq mi = 1 section
320 rd = 1 mi	160 acres = ¼ section
8,000 links = 1 mi	36 sq mi = 1 township (twp)
5,280 ft = 1 mi	6 mi sq = 1 twp
1,760 yd = 1 mi	1 sq mi = 2.59 sq km
9 in = 1 span	
4 in = 1 hand	

TABLE A-4

Volume Measurements

Cubic Measure
1,728 cu in = 1 cu ft
1 cu ft = 7.4805 gal (59.844 pt)
27 cu ft = 1 cu yd
128 cu ft (4′× 4′× 8′) = 1 cord (stacked wood)
231 cu in = 1 gal
2,150.4 cu in = 1 bushel (bu)
1.244 cu ft = 1 bu

(continued)

TABLE A-4 (Continued)

Liquid Measure

1 pt (4 gills)	=	16 fluid oz
1 qt (2 pt)	=	32 fluid oz
1 gal (4 qt)	=	128 fluid oz
1 gal (US)	=	0.8327 Imperial gal
31½ gal	=	1 barrel
42 gal	=	1 barrel (petroleum measure)
63 gal (2 barrels)	=	1 hogshead
1 acre-foot of water	=	325,851 gallons
1 acre-foot of water	=	1,233.5 cu m

Dry Measure

2 pints dry	=	1 quart dry
8 quarts dry	=	1 peck
4 pecks	=	1 bushel
105 quarts dry or 7,056 cu in	=	1 standard barrel

Gallons in Square or Oblong Tanks

To find the number of gallons in a square or oblong tank, multiply the number of cubic feet that it contains by 7.4805

Gallons in Circular Tanks

To find the number of gallons in a circular tank or well, square the diameter in feet, multiply by the depth in feet, and then multiply by 5.875

TABLE A-5
Calculating Rates of Application

1 acre-foot of soil	=	1,613 cu yd = 4,000,000 lb (approximate)
1 ton/acre	=	20.8 grams/sq ft
1 ton/acre	=	1 lb/21.78 sq ft
1 ton/acre	=	25.12 quintals/hectare
1 ton/acre-6 in depth of soil	=	1 gram/kilogram of soil
1 gram/sq ft	=	96 lb/acre

TABLE A-5 (Continued)

1 lb/acre = 0.0104 grams/sq ft
1 lb/acre = 1.121 kilograms/hectare
100 lb/acre = 2.296 lb/1,000 sq ft
grams/sq ft × 96 = lb/acre
kilograms/48 sq ft = tons/acre
(lb/sq ft) × 21.78 = tons/acre
(lb/sq ft) × 43,560 = lb/acre
(lb/1,000 sq ft) × 0.4878 = kilograms/100 sq m
(lb/cu yd) × 0.0005932 = grams/cubic centimeter
(lb/cu yd) × 0.5932 = kilograms/cubic meter
1,000 sq ft = 1/43.56 or 0.02296 acre
tons/acre-foot soil = 0.00136 × ppm
cu ft/second = 448.83 gal/minute
ppm = 17.1 × grains/gallon
ppm × 0.00136 = tons/acre-foot

TABLE A-6

Convenient Conversion Factors from A - Z

Multiply	By	To Get
Acres	0.4047	Hectares
Acres	43,560	Square feet
Acres	160	Square rods
Acres	4,840	Square yards
Bushels	4	Pecks
Bushels	64	Dry pints
Bushels	32	Dry quarts
Centimeters	0.3937	Inches
Centimeters	0.01	Meters
Cubic centimeters	0.03381	Ounces (liquid)
Cubic centimeters	0.06102	Cubic inches

(continued)

TABLE A-6 (Continued)

Cubic feet	1,728	Cubic inches
Cubic feet	0.03704	Cubic yards
Cubic feet	7.4805	Gallons
Cubic feet	29.92	Quarts (liquid)
Cubic yards	0.76455	Cubic meters
Cubic yards	27	Cubic feet
Cubic yards	46,656	Cubic inches
Cubic yards	202	Gallons
Feet	30.48	Centimeters
Feet	12	Inches
Feet	0.3048	Meters
Feet	0.060606	Rods
Feet	1/3 or 0.33333	Yards
Feet per minute	0.01136	Miles per hour
Gallons (U.S.)	0.1337	Cubic feet
Gallons (U.S.)	231	Cubic inches
Gallons (U.S.)	4	Quarts
Gallons (U.S.)	3.785	Liters
Gallons (Imperial)	4.546	Liters
Gallons (Imperial)	0.1605	Cubic feet
Gallons (Imperial)	277.4	Cubic inches
Gallons of water (US)	8.3436 at 0°C	Pounds of water
Gallons of water (US)	8.31217 at 25°C	Pounds of water
Grams	15.43	Grains
Grams	0.001	Kilograms
Grams	1,000	Milligrams
Grams	0.0353	Ounces
Grams per liter	1,000	Parts per million
Grams per cc	8.3443	Pounds per gallon
Grams per cc	1,685.55	Pounds per cubic yard

TABLE A-6 (Continued)

Hectares	2.471	Acres
Inches	2.54	Centimeters
Inches	0.08333	Feet
Kilograms	1,000	Grams
Kilograms	2.2046	Pounds
Kilograms per hectare	0.893	Pounds per acre
Megagrams per hectare	893	Tons per acre
Kilometers	3,281	Feet
Kilometers	0.6214	Miles
Liters	1,000	Cubic centimeters
Liters	0.001	Cubic meters
Liters	0.0353	Cubic feet
Liters	61.02	Cubic inches
Liters	0.2642	Gallons
Liters	1.057	Quarts (liquid)
Meters	100	Centimeters
Meters	3.2181	Feet
Meters	39.37	Inches
Miles	5,280	Feet
Miles	63,360	Inches
Miles	320	Rods
Miles	1,760	Yards
Miles per hour	88	Feet per minute
Miles per hour	1.467	Feet per second
Miles per minute	60	Miles per hour
Milliequivalents per liter (meq/L)	1	Millimoles per liter ($mmol_c\ L^{-1}$ or $mmol_c/L$)
Milliequivalents per 100 g (meq/100g)	Equivalent wt × 10	Parts per million

(continued)

TABLE A-6 (Continued)

Milliequivalents per 100g (meq/100g)	1	Centimoles per kilogram ($cmol_c\ kg^{-1}$ or $cmol_c/kg$)
Millimhos per centimeter (mmhos/cm)	1	Decisiemens per meter (x/m)
Millimhos per centimeter (mmhos/cm)	1,000	Micromhos per centimeter (μmhos/cm)
Ounces (dry)	0.0625	Pounds
Ounces (fluid-U.S.)	0.0625	Pints (liquid-U.S.)
Ounces (fluid-U.S.)	0.03125	Quarts (liquid-U.S.)
Ounces (fluid-U.S.)	29.574	Milliliters
Ounces (fluid-Imperial)	0.05	Pints (liquid-Imperial)
Ounces (fluid-Imperial)	28.410	Milliliters
Parts per million	8.345	Pounds per million gal water
Pecks	16	Pints (dry)
Pecks	8	Quarts (dry)
Pints (dry)	0.5	Quarts (dry)
Pints (liquid)	16	Ounces (liquid)
Pounds	453.5924	Grams
Pounds	16	Ounces
Pounds of water	0.1198	Gallons of water
Quarts (liquid)	0.9463	Liters
Quarts (liquid)	32	Ounces (liquid)
Quarts (liquid)	2	Pints (liquid)
Rods	16.5	Feet
Rods	5.5	Yards
Square feet	144	Square inches
Square feet	0.11111	Square yards
Square feet	0.09293	Square meters
Square inches	0.00694	Square feet
Square miles	640	Acres
Square miles	27,878,400	Square feet
Square rods	0.00625	Acres
Square rods	272.25	Square feet
Square yards	0.0002066	Acres

TABLE A-6 (Continued)

Square yards	9	Square feet
Square yards	1,296	Square inches
Temperature °C + 17.98	1.8	Temperature °F
Temperature °F − 32	5/9 or 0.5555	Temperature °C
Tons (short)	907.1849	Kilograms
Tons (short)	2,000	Pounds
Tons (long)	2,240	Pounds
Yards	3	Feet
Yards	36	Inches
Yards	0.9144	Meters

TABLE A-7

Zones of Equivalent Values

1 ppm = 1 mg/kg = 1 μg/g = 1 mg/L = 1,000 μg/L = 0.0001% = 1 lb/million lb
1% (w/v) = 10 g/L = 1.335 oz/gal = 8.34 lb/100 gal (U.S.)
1 g/cc = 8.34 lb/gal (U.S.)
1.60 oz/gal = 10.02 lb/100 gal (Imperial)
1% (v/v) = 1.28 fl oz/gal (U.S.) = 1 gal/100 gal
1 cu ft water = 7.48 gal (U.S.) = 62.5 lb = 28.32 L
1 mmhos/cm = 1,000 μmmhos/cm = 1 dS/m = 1 mS/cm = 1,000 μS/cm
1 meq/L = 1 $mmol_c$ L^{-1} = 1 mol m^{-3}
1 meq/100g = 1 $cmol_c$ kg^{-1} = 10 $mmol_c$/kg

TABLE A-8

Conversion of Acres to Smaller Units*

A. Conversion for Flower Pots or Flower Boxes

	Use These Volumes of Average Fertilizer (in tsp)				
	Flower Pots			Flower Boxes	
At These Rates per Acre	4-inch	6-inch	8-inch	1 Sq ft	2 Sq ft
500	0.05†	0.10†	0.30†	1	2
1,000	0.10†	0.20†	0.60†	2	4

(continued)

TABLE A-8 (Continued)

B. Conversion for Small Areas								
Find Your Material in the List Below	Find the Recommended Rate per Acre	Determine the Amount of Fertilizer Needed for Your Specific Area						
Material According to Approximate Weight per Pint	Amounts per Acre	Use These Weights per Specified Area		Use These Volumes per Specified Area				
						10 Feet of Row Spaced[‡]		
		100 Sq ft	1,000 Sq ft	10 Sq ft	100 Sq ft	1 ft	2 ft	3 ft
	lb	lb	lb	tbsp	pt	tbsp	tbsp	Cup
Activated sewage sludge Dried blood Sulfur (10 oz per pint)	100	0.2	2.3	1.2	0.4	1.2	2.4	0.2
	500	1.2	11.5	6	1.9	6	12	1.1
	1,000	2.3	23	12	3.7			
Ammonium chloride Ammonium nitrate Urea (13 oz per pint)	100	0.2	2.3	0.9	0.3	0.9	1.8	0.2
	500	1.2	11.5	4.5	1.4	4.5	9.0	0.8
	1,000	2.3	23	9	2.8			
Ammonium phosphate Gypsum Mixed fertilizers Potassium chloride (16 oz per pint)	100	0.2	2.3	0.7	0.2	0.7	1.4	0.1
	500	1.2	11.5	3.5	1.2	3.5	7	0.7
	1,000	2.3	23	7	2.3			
Ammonium sulfate Calcium nitrate Mixed fertilizers Superphosphate (19 oz per pint)	100	0.2	2.3	0.6	0.2	0.6	1.2	0.1
	500	1.2	11.5	3	1	3	6	0.6
	1,000	2.3	23	6	2			

TABLE A-8 (Continued)

Ground limestone	100 500	0.2 1.2	2.3 11.5	0.5 2.5	0.2 0.8	0.5 2.5	1 5	0.1 0.5
Potassium sulfate (23 oz per pint)	1,000 2,000	2.3 4.6	23 46	5 10	1.6 3.2			

*Reprinted from the U.C. Division of Ag and Natural Resources; originally printed as Leaflet No. 2285

†Since it is difficult to measure this small amount, you can dissolve 1 teaspoon of fertilizer in a pint of water and add the appropriate proportion to the flowerpot.

‡Since high rates are not desirable in row fertilization, they are omitted in this Table.

TABLE A-9
The Metric System

The fundamental unit of the metric system is the meter (the unit of length), from which the units of mass (gram) and capacity (liter) are derived; all other units are the decimal subdivisions or multiples thereof. These three units are simply related so that, for all practical purposes, the volume of 1 kilogram of water (1 liter) is equal to 1 cubic decimeter.

Prefix		Meaning	Units	
micro-	= one millionth	= 1/1,000,000	= 0.000001 (1×10^{-6})	
milli-	= one thousandth	= 1/1,000	= 0.001	
centi-	= one hundredth	= 1/100	= 0.01	
deci-	= one tenth	= 1/10	= 0.1	
unit	= one	= 1/1	= 1	■ meter for length ■ gram for mass ■ liter for capacity
deka- or deca-	= ten	= 10/1	= 10	
hecto-	= one hundred	= 100/1	= 100	
kilo-	= one thousand	= 1,000/1	= 1,000	
mega-	= one million	= 1,000,000/1	= 1,000,000 (1×10^{6})	

The metric terms are formed by combining the words "meter," "gram" and "liter" with the eight numerical prefixes. The finer subdivisions of measurement, nano- (10^{-9}), pico- (10^{-12}), and femto- (10^{-15}), are additional prefixes sometimes usefully employed, especially with units of mass (gram) and capacity (liter).

TABLE A-10
Metric – U.S. System Equivalents

Length

Metric Denominations and Values		U.S. Equivalents
myriameter	= 10,000 m	= 6.2137 mi
kilometer	= 1,000 m	= 0.62137 mi or 3,280 ft 10 in
hectometer	= 100 m	= 328 ft 1 in
dekameter	= 10 m	= 293.7 in
meter	= 1	= 39.37 in
decimeter	= 0.1 m	= 3.937 in
centimeter	= 0.01 m	= 0.3937 in
millimeter	= 0.001 m	= 0.0394 in

Volume

Name	Number of Liters	Cubic Measure	Dry Measure	Liquid Measure
kiloliter	1,000	= 1 cu m	= 1.308 cu yd	= 264.17 gal
hectoliter	100	= 0.1 cu m	= 2 bu 3.35 pk	= 26.417 gal
dekaliter	10	= 10 cu dm	= 9.08 qt	= 2.6417 gal
liter	1	= 1 cu dm	= 0.908 qt	= 1.0567 qt
deciliter	0.1	= 0.1 cu dm	= 6.1022 cu in	= 0.845 gill
centiliter	0.01	= 10 cu cm	= 0.6102 cu in	= 0.338 fl oz
milliliter	0.001	= 1 cu cm	= 0.061 cu in	= 0.27 fl dr

Weight

Name	Number of Grams	Cubic Measure[1]	Avoirdupois Wt
millier or tonneau	1,000,000	= 1 cu m	= 2,204.6 lb
quintal	100,000	= 1 hL	= 220.46 lb
myriagram	10,000	= 10 L	= 22.046 lb
kilogram	1,000	= 1 L	= 2.2046 lb
hectogram	100	= 1 dl	= 3.5274 oz
dekagram	10	= 10 cu cm	= 0.3527 oz

TABLE A-10 (Continued)

gram	1	= 1 cu cm	= 15.432 grains
decigram	0.1	= 0.1 cu cm	= 1.5432 gr
centigram	0.01	= 10 cu mm	= 0.15432 gr
milligram	0.001	= 1 cu mm	= 0.0154 gr

Area

Name	Metric	U.S. System
hectare	10,000 sq m	= 2.471 acres
are	100 sq m	= 119.6 sq yd
centare	1 sq m	= 1,550 sq in

[1] Based on pure water at 4°C and 760 mm pressure

TABLE A-11

Metric - U.S. System Conversions

To Convert Column 1 into Column 2, Multiply by	Column 1	Column 2	To Convert Column 2 into Column 1, Multiply by
Length			
0.621	kilometer, km	mile, mi	1.609
1.094	meter, m	yard, yd	0.914
0.394	centimeter, cm	inch, in	2.54
Area			
0.386	kilometer^2, km^2	mile^2, mi^2	2.59
247.1	kilometer^2, km^2	acre, ac	0.00405
2.471	hectare, ha	acre, ac	0.405
Volume			
0.00973	meter^3, m^3	acre-inch	102.8
3.532	hectoliter, hL	cubic foot, ft^3	0.2832
2.838	hectoliter, hL	bushel, bu	0.352
0.0284	liter, L	bushel, bu	35.24
1.057	liter, L	quart (fluid-US), qt	0.946

(*continued*)

TABLE A-11 (Continued)

Mass

1.1023	tonne (metric)	ton (English)	0.9072
2.2046	quintal, q	hundredweight, cwt (short)	0.4536
2.2046	kilogram, kg	pound, lb	0.4536
0.035	gram, g	ounce (avdp.), oz	28.35

Pressure

100	bar	kilopascal, kPa	0.01
10	bar	megapascal, mPa	0.1
14.5038	bar	lb/in^2, psi	0.06895
0.9869	bar	atmosphere, atm	1.013
0.9678	kg (weight)/cm^2	atmosphere, atm	1.033
14.22	kg (weight)/cm^2	lb/in^2, psi	0.07031
14.70	atmosphere, atm	lb/in^2, psi	0.06805
0.1450	kilopascal	lb/in^2, psi	6.895
0.009869	kilopascal	atmosphere, atm	101.3

Yield or Rate

0.4461	tonne (metric)/ hectare	ton (U.S.)/acre	2.2416
0.3983	tonne (metric)/ hectare	ton (British)/acre	2.5106
0.8922	kilogram/hectare	lb/ac	1.1208
0.8922	quintal/hectare	hundredweight/ acre	1.1208
1.15	hectoliter/hectare	bu/acre	0.87

TABLE A-12

Nursery Container Sizes and Number of Containers per Cubic Yard

Container Sizes	Approximate Number of Containers per Cubic Yard
5″ standard	800
6″ standard	450
1 gallon	300

TABLE A-12 (Continued)

2 gallon	140
3 gallon	80
5 gallon	50
7 gallon	35

Large Container Measurements

Container Sizes	Surface Area in Sq Ft
10 gallon	1.4
15 gallon	1.5
20 gallon	2.3
25 gallon	2.8
45 gallon	4.8
24 inch box	4.0
30 inch box	6.25
36 inch box	9.0
48 inch box	16

TABLE A-13

Number of Trees or Plants per Unit Area

	Number per			Number per	
Spacing	Acre	1,000 sq ft	Spacing	Acre	1,000 sq ft
1 by 2 ft	21,780	500	6 by 6 ft	1,210	28
1 by 3 ft	14,520	333	6 by 8 ft	907	21
1 by 4 ft	10,890	250	8 by 8 ft	680	16
1½ by 2 ft	14,520	333	10 by 10 ft	436	10
1½ by 3 ft	9,680	222	12 by 12 ft	302	
2 by 3 ft	7,260	167	15 by 15 ft	194	
2 by 4 ft	5,445	125	16 by 16 ft	170	
3 by 4 ft	3,630	83	18 by 18 ft	134	
3 by 5 ft	2,904	67	20 by 20 ft	109	
3 by 6 ft	2,420	56	25 by 25 ft	70	
4 by 4 ft	2,722	62	30 by 30 ft	48	
4 by 6 ft	1,815	42	40 by 40 ft	27	

TABLE A-14

Temperature Comparison of Celsius to Fahrenheit

Celsius (C°)	Fahrenheit (F°)
–30	–22
–20	–4
–10	14
0	32
10	50
20	68
30	86
40	104
50	122
60	140
70	158
80	176
90	194
100	212

Conversion Formulas

°C = 5/9 (°F – 32)	°F = (9/5 °C) + 32

TABLE A-15

Atomic Weights, and Common Valence Values

Name	Symbol	Atomic Weight	Common Valence
Aluminum	Al	26.98	3
Boron	B	10.81	3
Calcium	Ca	40.08	2
Carbon	C	12.01	–4, 4
Chlorine	Cl	35.45	–1
Cobalt	Co	58.93	2
Copper	Cu	63.55	2

TABLE A-15 (Continued)

Fluorine	F	19.00	–1
Hydrogen	H	1.01	–1, 1
Iodine	I	126.90	–1
Iron	Fe	55.85	2, 3
Magnesium	Mg	24.31	2
Manganese	Mn	54.94	2, 4
Molybdenum	Mo	95.94	2, 6
Nickel	Ni	58.69	2
Nitrogen	N	14.01	–3, 5
Oxygen	O	16.00	–2
Phosphorus	P	30.97	–3,5
Potassium	K	39.10	1
Sodium	Na	23.00	1
Sulfur	S	32.07	–2, 6
Zinc	Zn	65.38	2

TABLE A-16

Chemical Symbols, Equivalent Weights, and Common Names of Ions, Salts, and Chemical Amendments

Chemical Symbol or Formula	Gram Equivalent Weight	Common Name
Ca^{2+}	20.04	Calcium ion
Mg^{2+}	12.15	Magnesium ion
Na^{+}	23.00	Sodium ion
K^{+}	39.10	Potassium ion
Cl^{-}	35.46	Chloride ion
NO_3^{-}	62.01	Nitrate ion
NH_4^{+}	17.03	Ammonium ion
SO_4^{2-}	48.03	Sulfate ion
CO_3^{2-}	30.00	Carbonate ion
HCO_3^{-}	61.02	Bicarbonate ion
$CaCl_2$	55.50	Calcium chloride

(*continued*)

TABLE A-16 (Continued)

$CaSO_4$	68.07	Calcium sulfate
$CaSO_4 \cdot 2H_2O$	86.09	Gypsum
$CaCO_3$	50.04	Calcium carbonate
$MgCl_2$	47.62	Magnesium chloride
$MgSO_4$	60.19	Magnesium sulfate
$MgCO_3$	42.16	Magnesium carbonate
NaCl	58.46	Sodium chloride
Na_2SO_4	71.03	Sodium sulfate
Na_2CO_3	53.00	Sodium carbonate
$NaHCO_3$	84.02	Sodium bicarbonate
KCl	74.56	Potassium chloride
K_2SO_4	87.13	Potassium sulfate
K_2CO_3	69.10	Potassium carbonate
$KHCO_3$	100.12	Potassium bicarbonate
S	16.03	Sulfur
SO_2	32.03	Sulfur dioxide
H_2SO_4	49.04	Sulfuric acid
$Al_2(SO_4)_3 \cdot 18H_2O$	111.08	Aluminum sulfate
$FeSO_4 \cdot 7H_2O$	139.02	Iron sulfate (ferrous)

TABLE A-17

Salt Index (Relative Effect of Fertilizer Materials on the Soil Solution)[1]

	% Analysis						
Material	**N**	**P_2O_5**	**K_2O**	**S**	**Ca**	**Index**	**Partial**
Anhydrous ammonia	82					47.1	0.572
Ammonium nitrate	34					104.7	2.990
Ammonium phosphate	11	48				26.9	2.442
Ammonium polysulfide	20			40		43.6	2.180
Ammonium sulfate	21			22		69.0	3.253
Ammonium thiosulfate	12			26		84.4	7.040
Limestone (calcium carbonate)					35	4.7	0.083
Calcium cyanamide	21				37	31.0	1.476
Calcium ammonium nitrate	26				10	61.1	2.982
Calcium nitrate	15				19	52.5	4.409

TABLE A-17 (Continued)

Ca sulfate dihydrate gypsum				17	23	8.1	0.247
Di-ammonium phosphate	18	48				29.9	1.614
Dolomite					20	0.8	0.042
Manure salts, 20%						113.0	5.636
Manure salts, 30%						91.9	3.067
Mono-ammonium phosphate	11	52				34.2	2.453
Mono-calcium phosphate		20			40	15.4	0.274
Sodium nitrate	16					100.0	6.060
Potassium chloride 50			50			109.0	2.189
Potassium chloride 60			60			116.0	1.936
Potassium chloride 63			63			114.0	1.812
Potassium nitrate	13		45			73.6	5.336
Potassium sulfate			50	18		46.1	0.853
Potassium thiosulfate			25	17		64.0	2.560
Mono-potassium phosphate		52	34			8.4	0.097
Sodium chloride						154.0	2.899
Potassium magnesium sulfate			22	22		43.2	1.971
Superphosphate, 16%		16				7.8	0.487
Superphosphate, 20%		20				7.8	0.390
Superphosphate, 45%		45				10.1	0.224
Superphosphate, 48%		48				10.1	0.210
UAN 32%	32					95.0	2.304
Urea	46					75.4	1.618

[1]After L.F. Rader, Jr., et al., Soil Sci., 55:21–218, 1943

TABLE A-18

Nutrient Analyses of Some Organic Materials

Description	%N	%P_2O_5	%K_2O
Non-Composted Poultry (dry)			
Turkey/rice hull litter	1.75	2.65	1.85
Fresh broiler/rice huller	3.90	2.55	2.65
Fresh layer	3.95	6.25	3.35
Aged layer	2.15	8.20	3.95

(continued)

TABLE A-18 (Continued)

Non-Composted Dairy/Steer (dry)			
Fresh dairy separator solids	2.15	0.85	0.60
Fresh dairy corral scrapings	2.35	1.30	0.70
Aged dairy separator solids	2.05	0.65	0.40
Aged dairy corral scrapings	1.30	1.55	3.30
Composts (dry)			
Broiler/rice hull compost	1.90	4.30	2.50
Dairy	1.35	1.35	2.85
Dairy/gin trash	1.55	1.10	2.85
Dairy/steer	1.65	0.85	2.55
Dairy/poultry	1.70	1.95	3.30
Gin trash	2.25	0.80	2.50
Vermicompost (earthworm)	1–3.4	0.07–0.25	0.77–1.41
Bulky Organic Materials (dry)			
Alfalfa hay	2.50	0.50	2.10
Bean straw	1.20	0.25	1.25
Grain straw	0.60	0.20	1.10
Cotton gin trash	1.65	0.55	1.90
Seaweed (kelp)	0.20	0.10	0.60
Winery pommace	1.50	1.50	0.75
Concentrated Organic Materials (as received)			
Alfalfa meal	2–3	0.5–1	1–3
Cottonseed meal	6–7	0.4–3	1–2
Blood meal	10–14	1–2	0.6–1
Bone meal	0–2	10–14	
Hoof and horn meal	10–14	1–2	
Fish meal	5–12	3–8	2–4
Fish emulsion	5–6	1–2	1–2
Kelp meal	1–1.5	0.1–0.5	2–4
Feather meal	12		
Bat guano	8–13	2–5	0.05–2

TABLE A-18 (Continued)

Sewage sludge (dried biosolids)	0.1–17.6	0.1–14.3	0.1–2.6
Tankage (dried animal residue)	7	8.6–10	0.5–1.5

Notes:

- Content of organic materials varies widely, according to source and method of processing. Refer to "Organic Amendments and Fertilizers" Publication #21505, University of California.
- Not all of the listed materials are approved for organic production. Check with your certifying agency.
- Registered organic fertilizers generally have their content expressed as total nitrogen, available phosphate, and soluble potash (as received) and not as total nitrogen, phosphate and potash (dry weight basis) as with manures, composts and other bulky materials. Always read the label.

TABLE A-19

Nitrogen Fertilizer Solubilities at Various Temperatures lbs of Dry Fertilizer/1,000 lb Water

Temp (°F)	KNO_3	Urea	AN	UAN-32	AS	MAP	DAP
32°	130	670	1,180	3,900	706	227	429
35°	143	698	1,246	4,552		238	462
40°	166	745	1,357	5,639	718	257	517
45°	188	793	1,479	6,725		276	572
50°	210	840	1,580	7,812	730	295	628
55°	240	898	1,683	8,900		317	645
60°	271	956	1,788	9,987		339	662
65°	301	1,020	1,888	11,074		361	679
70°	335	1,073	1,991	12,161	749	384	696
75°	374	1,153	2,150	13,247		409	713
80°	403	1,233	2,259	14,333		434	730
85°	452	1,314	2,393	15,420	777	459	747

KNO_3 = potassium nitrate; AN = ammonium nitrate; UAN-32 = urea ammonium nitrate 32%; AS = ammonium sulfate; MAP = monoammonium phosphate; DAP = diammonium phosphate

TABLE A-20

Nitrogen Fertilizer Percent Analysis of Saturated Solution at Various Temperatures

Temp (°F)	Urea–N	AN–N	UAN–32-N	AS–N	AS–S	MAP–N	MAP–P_2O_5	DAP–N	DAP–P_2O_5
32°	18.7	19.0	32.6	8.8	10.0	2.3	11.4	6.4	16.2
35°	19.2	19.4	33.6			2.3	11.9	6.7	17.3
40°	19.9	20.2	34.8			2.5	12.6	7.2	18.0
45°	20.6	20.8	35.7			2.6	13.0	7.7	19.6
50°	21.3	21.4	36.3	8.8	10.2	2.8	13.8	8.2	20.6
55°	22.1	22.0	36.8			2.9	14.9	6.3	21.1
60°	22.6	22.4	37.2			3.1	15.6	6.4	21.4
65°	23.5	22.9	37.6			3.2	16.4	8.6	21.7
70°	24.2	23.4	37.9	9.1	10.4	3.4	17.1	8.7	22.1
75°	25.0	23.8	38.1			3.5	17.9	8.8	22.3
80°	25.8	24.3	38.3			3.7	18.7	8.9	22.7
85°	26.5	24.7	38.5	9.3	10.6	3.8	19.4	9.1	22.9

AN–N = ammonium nitrate-nitrogen; UAN–32–N = urea ammonium nitrate-nitrogen; AS–N = ammonium sulfate-nitrogen; AS–S = ammonium sulfate-sulfur; MAP–N = monoammonium phosphate-nitrogen; MAP–P_2O_5 = monoammonium phosphate-phosphate; DAP–N = diammonium phosphate-nitrogen; DAP–P_2O_5 = diammonium phosphate-phosphate

TABLE A-21

Potassium Fertilizer Solubilities at Various Temperatures lb of Dry Fertilizer/1,000 lb Water

Temp (°F)	KCl	K_2SO_4	MKP	DKP
32°	280	74	142	1,328
35°	285			
40°	293		158	
45°	302			
50°	310	92	178	1,488
55°	318			
60°	327	115	202	1,513

TABLE A-21 (Continued)

65°	335			
70°	343		229	1,605
75°	352		249	1,677
80°	360		255	1,724
85°	368	128	274	1,788

KCl = potassium chloride; K_2SO_4 = potassium sulfate; MKP = monopotassium phosphate; DKP = dipotassium phosphate

TABLE A-22

Potassium Fertilizer Percent Analysis of Saturated Solution at Various Temperatures

Temp (°F)	KN–K_2O	KN–N	KCl–K_2O	KCl–Cl_2	KS–K_2O	KS–S	MKP–K_2O	MKP–P_2O_5	DKP–K_2O	DKP–P_2O_5
32°	5.4	1.6	13.6	10.4	3.7	1.3	4.3	6.4	30.8	23.3
35°	5.8	1.7	14.0	10.5						
40°	6.6	2.0	14.3	10.6			4.7	7.0		
45°	7.4	2.2	14.7	11.0						
50°	8.1	2.4	14.9	11.3	4.6	1.6	5.2	7.7	32.0	24.1
55°	9.0	2.7	15.2	11.5						
60°	9.9	3.0	15.6	11.7	5.6	1.9	5.9	8.6	32.6	24.5
65°	10.8	3.2	15.9	11.9						
70°	11.7	3.5	16.1	12.1			6.4	9.6	33.3	25.1
75°	12.7	3.8	16.4	12.4			6.7	10.0	33.6	25.6
80°	13.4	4.0	16.7	12.6			7.0	10.5	34.2	25.8
85°	14.5	4.3	17.0	12.8	6.1	2.1	7.4	11.0	34.7	26.1

KN–K_2O = potassium nitrate-potash; KN–N = potassium nitrate-nitrogen; KCl–K_2O = potassium chloride-potash; KCl–Cl_2 = potassium chloride-chloride; KS–K_2O = potassium sulfate-potash; KS–S = potassium sulfate-sulfur; MKP–K_2O = monopotassium phosphate-potash; MKP–P_2O_5 = monopotassium phosphate-phosphate; DKP–K_2O = dipotassium phosphate-potash; DKP–P_2O_5 = dipotassium phosphate-phosphate

APPENDIX B

How to Measure Areas

WHAT IS AREA?

The area of an object can be described as the number of unit squares that can fit into a two dimensional surface. The symbol for area is usually designed as "A". Areas are measured in unit squares, for example square feet, square miles, etc. The known area of a particular landscape is important so that proper quantities of fertilizers, amendments, or other materials can be calculated and applied. Most landscapes consist of irregular shapes and sizes; therefore, the standard formulas for determining areas cannot always be used. This appendix provides methods for determining the area of regular and irregular shaped landscape surfaces.

CALCULATING AREAS OF REGULAR OBJECTS

Square

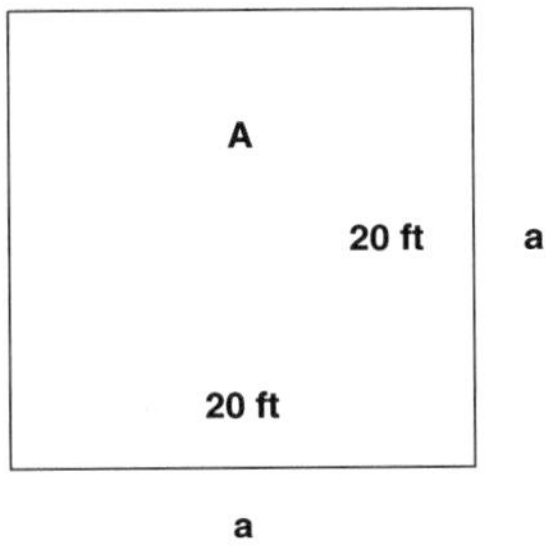

Formula: A = a × a
A = length and width

Example: a = 20 ft
A = 20 ft × 20 ft = 400 sq ft

Rectangle

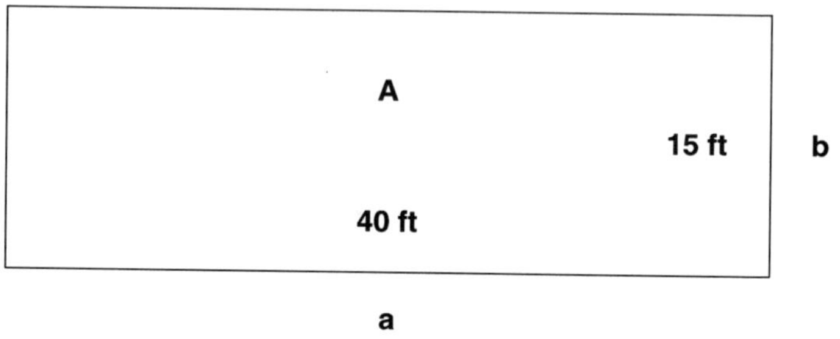

Formula: $A = a \times b$

a = length; b = width

Example: $a = 40$ ft; $b = 15$ ft

$A = 40\,\text{ft} \times 15\,\text{ft} = 600$ sq ft

Right Triangle

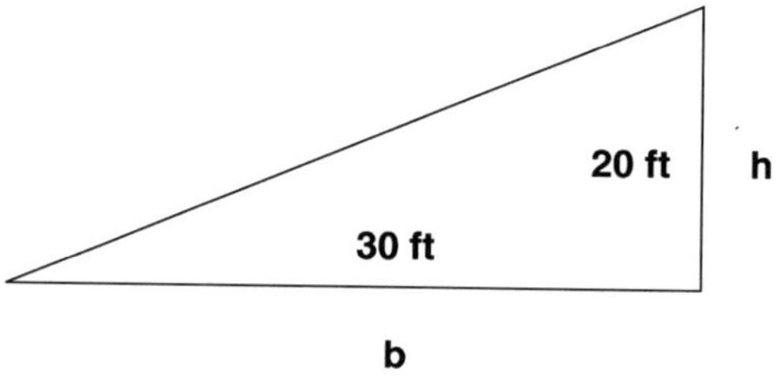

Formula: $A = \dfrac{b \times h}{2}$

b = base; h = height

Example: $b = 30$ ft; h = height 20 ft

$$A = \frac{20\,\text{ft} \times 30\,\text{ft}}{2} = 300\,\text{sq ft}$$

Circle

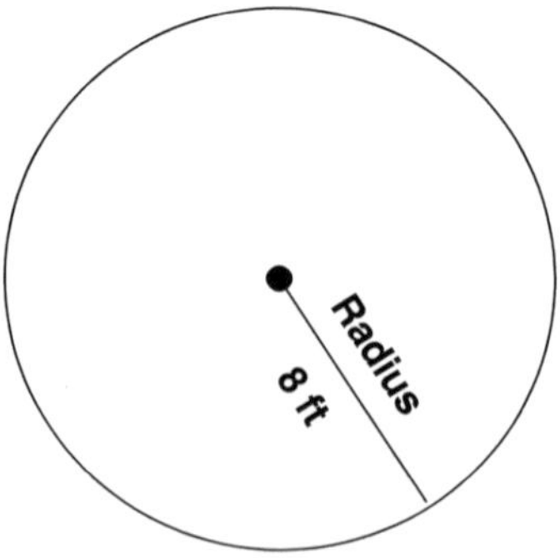

Formula: $A = \pi r^2$

$\pi = 3.14$

r = radius

Example: $\pi = 3.14$; $r = 8\,\text{ft}$

$A = 3.14 \times 8\,\text{ft} \times 8\,\text{ft}$

$= 200.96$ sq ft (round to 201)

Ellipse

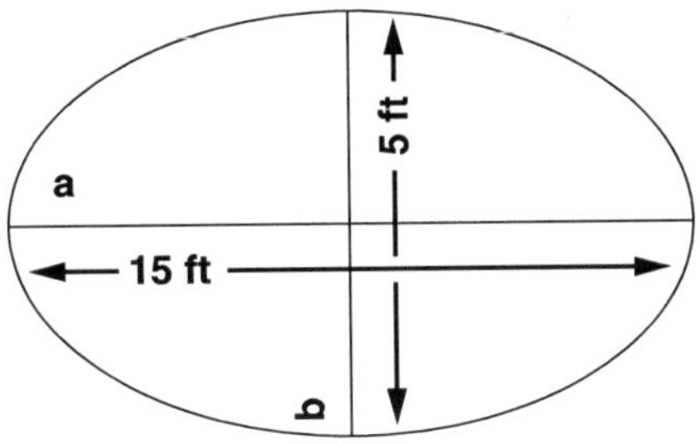

Formula: $A = 0.785 \times a \times b$

a = length of the ellipse

b = shorter dimension (width)

Example: $a = 15\,ft$; $b = 5\,ft$

$A = 0.785 \times 15\,ft \times 5\,ft$

$= 58.9\,sq\,ft$

CALCULATING AREAS OF IRREGULAR OBJECTS

Online and Electronic Mapping Tools

Today's electronic tools make determining the area of a space easy. Many online mapping tools, such as Google™ Earth, can calculate areas based on satellite images. Global Positioning System (GPS) devices are able to assist growers and advisers in precisely calculating landscape and turf areas. Horticulturalists are encouraged to take advantage of these systems.

Using Regular Geometric Shapes to Determine Irregular Shaped Areas

Determining the size of an irregular shaped area, such as a golf course, can be simplified by dividing it into regular geometric shapes, determining the areas of each, and then summing the area totals. Generally, any area can be considered a square, rectangle, triangle, circle or ellipse. For example, the fairways of a golf course can be visualized as rectangles, its tees as squares, and its greens, lakes, and water reservoirs as circles. One can choose to measure the dimensions of the regular geometric shapes using a measuring tape, a measuring wheel, or pacing. A pace is commonly assumed to be 3 feet in length. The following examples show how to determine the area of two irregular shaped landscapes.

Stepping Off and Calculating Approximate Areas

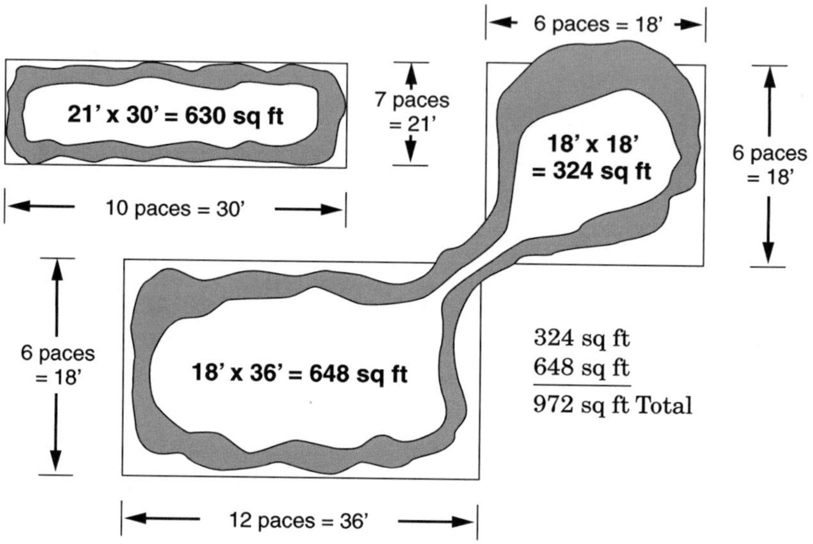

Note: A pace is approximately 3 feet.

Irregular Shape

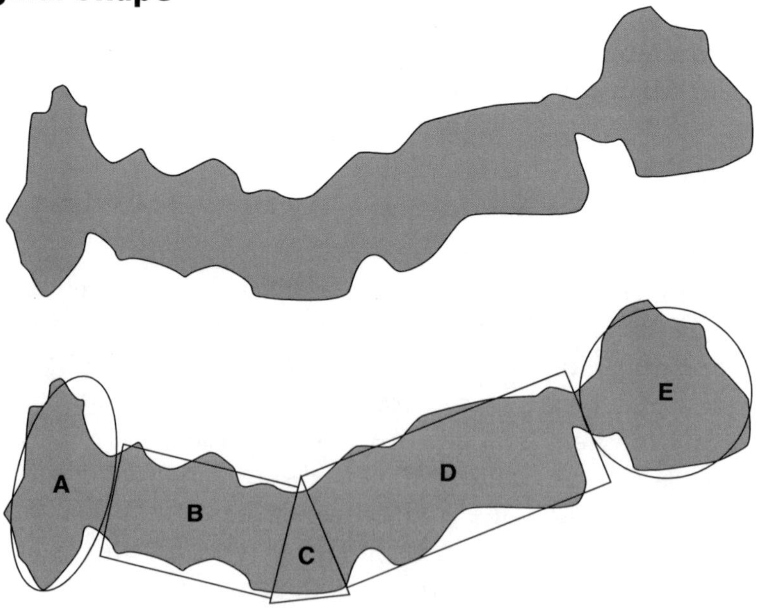

An irregular area reduced to simple geometric shapes

Area total = $A_A + A_B + A_C + A_D + A_E$

Determining the Area of Extremely Irregular Landscapes

Landscapes are rarely regular in shape. The area of a very irregular landscape can be obtained using the following method. First, find the longest line segment through the center of the area, as in line segment AB below. Second, draw or pace several lines perpendicularly to the center line. The total number of perpendicular lines will depend upon the irregularity of the shape; the more irregular, the more lines that should be drawn. Determine the average length of the perpendicular lines and use this value as the width in the formula for a rectangular area.

Formula: $A = a \times b$

a = distance between A and B

b = average of all 10 lengths a′ to j′ (lines are drawn perpendicular to AB)

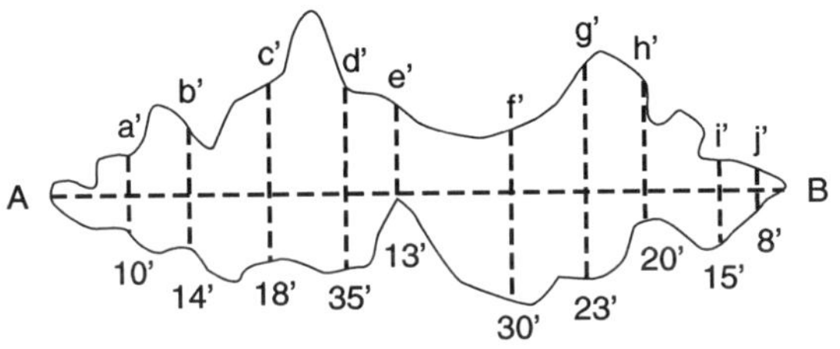

Example:

a′ = 10 ft; b′ = 14 ft; c′ = 18 ft; d′ = 35 ft; e′ = 13 ft; f′ = 30 ft; g′ = 23 ft; h′ = 20 ft; i′ = 15 ft; j′ = 8 ft

Sum of 10 lines a′ through j′ = 186 ft

AB = 128 ft

$b = 186 \div 10 = 18.6$ ft

$A = 128 \times 18.6 = 2{,}380$ sq ft

Determining the Area of An Irregular Shaped Circular Landscape

Another method for determining the size of an irregularly-shaped area, a golf green for example, is to consider whether it is almost circular. If so, establish a point as near to the center of the area as can be estimated. From this point, calculate 10-degree increments, as with a compass, at the edge of the circular area.

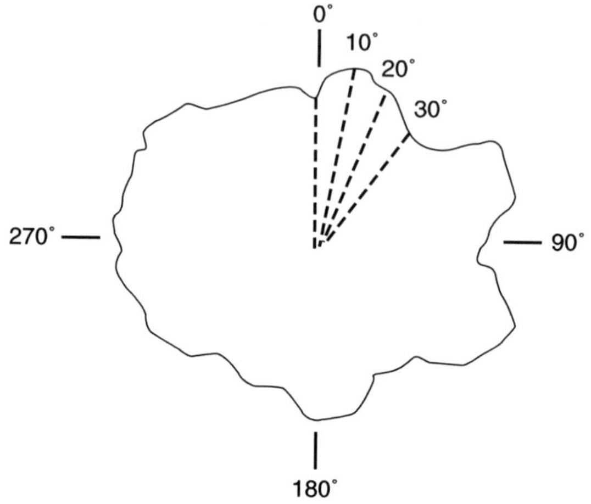

Then, measure the radius 36 times around the circular area and determine the average radius. Use the average radius to compute the area of the circle.

Formula: $A = \pi r^2$

Example:

If the sum of 36 radii is 1,980 ft, then the average radius = 1,980 ft ÷ 36 = 55 ft

$A = 3.14 \times (55\,\text{ft})^2 = 9{,}498.5$ sq ft (round to 9,500)

Appendix C

Related Professional Organizations

Agricultural Retailers Association, 1156 15th Street NW, Suite 302, Washington, DC 20005; *www.aradc.org*

American Horticultural Society, 7931 East Boulevard Drive, Alexandria, VA 22308; *www.ahs.org*

American Society for Horticultural Science, 1018 Duke Street, Alexandria, VA 22314; *www.ashs.org*

American Society of Agronomy, 5585 Guilford Road, Madison, WI 53711-5801; *www.agronomy.org*

American Society of Golf Course Architects, 125 North Executive Drive, Suite 302, Brookfield, WI 53005; *www.asgca.org*

American Society of Landscape Architects, 636 Eye Street NW, Washington, DC 2001-3736; *www.asla.org*

Arizona Crop Protection Association, 1110 East Missouri Avenue, Suite 340, Phoenix AZ, 85014; *www.azcropprotection.com*

Association of American Plant Food Control Officials, *www.aapfco.org*

Betrock International Horticultural Network, 7770 Davie Road Extension, Hollywood, FL 33024; *www.plantfinder.com*

California Association of Nurseries and Garden Centers, 1521 I Street Sacramento, CA 95814; *www.cangc.org*

California Association of Pest Control Advisers, 2300 River Plaza Drive, Suite 120, Sacramento, CA 95833; *www.capca.com*

California Certified Crop Adviser Program, 1143 North Market Boulevard, Suite 7, Sacramento, CA 95834; *www.cacca.org*

California Fertilizer Foundation, 4460 Duckhorn Drive, Suite A, Sacramento, CA 95834; *www.calfertilizer.org*

California Landscape Contractors Association, 1491 River Park Drive, Suite 100, Sacramento, CA 95815; *www.clca.org*

Canadian Fertilizer Institute, 350 Sparks Street, Suite 802, Ottawa, Ontario, K1R 7S8, Canada; *www.cfi.ca*

Certified Crop Adviser, 5585 Guilford Road, Madison, WI 53711-5801; *www.certifiedcropadviser.org*

Croplife of America, 1156 15th Street NW, Washington, DC 20005; *www.croplifeamerica.org*

Far West Agribusiness Association, 111 East Magnesium Road, Suite C, Spokane, WA 99208; *www.fwaa.org*

Fluid Fertilizer Foundation, 2805 Claflin Road, Suite 200, Manhattan, KS 66502; *www.fluidfertilizer.com*

Golf Course Builders Association, 727 O Street, Lincoln, NE 68508; *www.gcbaa.org*

Golf Course Superintendents Association of America, 1421 Research Park Drive, Lawrence, KS 66049-3859; *www.gcsaa.org*

International Plant Nutrition Institute, 3500 Parkway Lane, Suite 550, Norcross, GA 30092; *www.ipni.net*

International Society of Arboriculture, PO Box 3129, Champaign, IL 61826; *www.isa-arbor.com*

International Society of Horticultural Sciences, PO Box 500, 3001 Leuven 1, Belgium; *www.ishs.org*

North Carolina State University Turf Files, 2101 Hillsborough Street, Raleigh, NC 27695-7001; *www.turffiles.ncsu.edu*

Ohio State University Horticulture and Crop Science, 202 Kottman Hall, 2021 Coffey Road, Columbus, OH 43210; *www.hcs.osu.edu*

Pesticide Applicators Professional Association, PO Box 80095, Salinas, CA 93912-0095; *www.papaseminars.com*

Professional Landcare Network, 950 Herndon Parkway, Suite 450, Herndon, VA 20170; *www.landcarenetwork.org*

Responsible Industry for a Sound Environment, 1156 15th Street NW, Suite 400, Washington, D.C. 20005; *www.pestfacts.org*

Rocky Mountain Agribusiness Association, 3030 West 81st Avenue, Westminster, CO 80031; *www.rmagbiz.org*

Soil Science Society of America, 5585 Guilford Road, Madison, WI 53711-5801; *www.soils.org*

Southern California Turfgrass Council, 5215 State Street, Montclair, CA 91763; *www.turfcouncil.com/NewSite/mission.html*

Southwest Trees and Turf, PO Box 12507, Las Vegas, NV 89112-0507; *www.swtreesandturf.com*

Sports Turf Managers Association, 805 New Hampshire, Suite E, Lawrence, KS 66044; *www.stma.org*

The Fertilizer Institute, 425 Third St. SW, Suite 950, Washington, D.C. 20024; *www.tfi.org*

The Sulphur Institute, 1140 Connecticut Ave. NW, Suite 612, Washington, DC 20036; *www.sulphurinstitute.org*

Turf Net, PO Box 476, Middlebury, VT 05753; *www.turfnet.com*

Turfgrass Science at University of Minnesota, 1970 Folwell Avenue, 254 Alderman Hall, St. Paul, MN 55108; *www.turf.umn.edu*

United States Golf Association, PO Box 708, Far Hills, NJ 07931; *www.usga.org*

Western Plant Health Association, 4460 Duckhorn Drive, Suite A, Sacramento, CA 95834; *www.healthyplants.org*

Glossary of Terms

AAPFCO—American Association of Plant Food Control Officials.

ABSORPTION—The process by which a substance is taken into and included within another substance, e.g., intake of water by soil or intake of gases, water, nutrients, or other substances by plants. Not to be confused with adsorption.

ACID—A condition of an aqueous solution which has a greater concentration of hydrogen ions than hydroxyl ions.

ACID-FORMING—A term applied to any fertilizer that tends to make the soil more acidic.

ACID SOIL—A soil with a pH value below 7.0. A soil having relatively more hydrogen ions than hydroxyl ions in the soil solution.

ACTINOMYCETES—A class of microbes related to bacteria that form fungal-like structures.

ACTIVATED SEWAGE SLUDGE—An organic fertilizer made from sewage freed from grit and coarse solids and aerated after being inoculated with microorganisms. The resulting flocculated organic matter is withdrawn from the tanks, filtered with or without the aid of coagulants, dried, ground, and screened.

ACTIVE ACIDITY—A measure of hydrogen ions present in the soil solution. Active acidity is in equilibrium with potential (exchangeable) acidity.

ACTIVITY INDEX—The percentage of Cold Water Insoluble Nitrogen (CWIN) that is soluble in hot (100° C) water, of not less than 40 percent.

ADSORPTION—The increased concentration of molecules or ions at a surface, including exchangeable cations and anions on soil particles. See SODIUM ADSORPTION RATIO.

AEOLIAN—Wind-deposited material.

AERATION, SOIL—The exchange of air in soil with air from the atmosphere. The composition of the air in a well-aerated soil is similar to the composition of the atmosphere; the air in a poorly aerated soil is considerably higher in carbon dioxide and lower in oxygen than the air in the atmosphere above the soil.

AEROBIC—(1) Having molecular oxygen as part of the environment. (2) Growing only in the presence of molecular oxygen (such as aerobic organisms).

AEROPONICS—A hydroponic method where plant roots are suspended in an enclosed chamber and sprayed either continuously or intermittently with a nutrient solution.

AGGREGATE—A group of soil particles cohering so as to behave mechanically as a unit.

AGRICULTURAL CHEMICALS—Synthetic or naturally occurring compounds used in agriculture to control pests. These chemicals include algaecides, biocides, bactericides, fungicides, herbicides, insecticides, nematicides, and rodenticides.

AIR-FILLED POROSITY—The volume percentage of the medium occupied by air.

ALKALINE—A basic reaction in which the pH reading is above 7.0, as distinguished from acidic reaction in which the pH reading is below 7.0.

ALKALINE SOIL—A soil that has an alkaline reaction, i.e., a soil for which the pH reading of the saturated soil paste is above 7.0.

ALKALINITY—The term used when referring to the combined concentrations of bicarbonate and carbonate ions in water.

ALKALI SOIL—See SODIC SOIL.

ALLELOPATHIC COMPOUNDS—Chemicals produced by some plants to limit the growth of other plants.

ALLUVIAL—When water is the transporting agent for parent materials by stream.

AMENDMENT—Any material, such as lime, gypsum, manure, compost, sawdust, vermiculite, or a synthetic conditioner, that is worked into the soil to make it more productive. Strictly, a fertilizer is also an amendment, but the term amendment is used more commonly for added materials other than fertilizer.

AMINIZATION—A process in which proteins and N-rich compounds are broken down into amino acids.

AMINO ACIDS—Nitrogen-containing organic compounds, large numbers of which link together in the formation of the protein molecule. Each amino acid molecule contains one or more amino ($-NH_2$) groups and at least one carboxyl ($-COOH$) group. In addition, some amino acids (cystine, cysteine, and methionine) contain sulfur.

AMMONIATION—A process wherein ammonia (anhydrous, aqua, or a solution containing ammonia and other forms of nitrogen) is used to treat superphosphate to form ammoniated superphosphate, or to treat a mixture of fertilizer ingredients (including various acids) in the manufacture of multinutrient fertilizer.

AMMONIFICATION—Formation of ammonium compounds or ammonia.

ANAEROBIC—(1) The absence of molecular oxygen. (2) Growing in the absence of molecular oxygen, such as anaerobic bacteria.

ANALYSIS—The percentage composition as found by chemical analysis, expressed in those terms that the law requires and permits. Although *analysis* and *grade* sometimes are used synonymously, the term *grade* is applied only to the three primary plant foods—nitrogen (N), phosphate (P_2O_5), and potash (K_2O)—and is stated as the guaranteed minimum quantities present. (See also GRADE.)

ANGLE OF REPOSE—The angle between the horizontal and the slope of a pile of loose material at equilibrium.

ANHYDROUS AMMONIA—A gaseous end product of the high temperature and pressure, catalytic reaction between nitrogen and hydrogen gasses. The starting material for all nitrogenous fertilizers. Can be liquified and applied as a fertilizer or reacted to form other fertilizers.

ANION—An ion carrying a negative electrical charge. Examples include bicarbonate (HCO_3^-), carbonate (CO_3^{2-}), Cl^-, and SO_4^{2-}.

ANNUAL—A plant that completes its life cycle in one year.

ANTI-CRUSTANT—A material that when applied to the soil surface will have the property of either stabilizing soil aggregates or dissolving calcareous layers, thereby alleviating the adverse effects of raindrop or sprinkler impact and improving water penetration. Polyacrylamides (PAM) and acids are examples.

AOAC—Association of Official Analytical Chemists (of North America).

APATITE (Rock Phosphate)—A mineral phosphate having the type formula Ca10(X2)(PO4)6, where X is usually fluorine, chlorine, or the hydroxyl group, either singly or together.

AQUA AMMONIA (AQUA)—Anhydrous ammonia dissolved in water.

ATTAPULGITE CLAY—A gelling clay used in suspension fertilizer.

AUTOTROPHIC BACTERIA—Single celled organisms capable of utilizing carbon dioxide as a carbon source for growth.

AVAILABLE—In general, a form of nutrient capable of being assimilated by a growing plant. Available nitrogen is the nitrogen that is water soluble plus what can be made soluble or converted into free ammonia. Available phosphate is that portion that is water soluble plus the part that is soluble in ammonium citrate. Available potash is that portion soluble in water or in a solution of ammonium oxalate.

AVAILABLE NUTRIENT IN SOILS—The part of the supply of a plant nutrient in soils that can be taken up by plants at rates and in amounts significant to plant growth.

AVAILABLE WATER CAPACITY—The part of the water in soils that can be taken up by plants at rates significant to their growth; water that is usable or obtainable; water held between field moisture capacity (approximately 1/3 atmosphere) and permanent wilting point (15 atmospheres).

BACKFILL—Soil amendments mixed with the soil removed from the planting hole when planting landscape shrubs and trees.

BANDED FERTILIZER—Placement of fertilizer in a concentrated zone either on or below the soil surface.

BANDING—Method of fertilizer application. Banding is a general term for applications that concentrate fertilizers into narrow zones that are kept intact to provide a concentrated source of nutrients. Applications may be made prior to, during, or after planting.

BASE EXCHANGE—The replacement of basic cations (Ca, Mg, Na, and K), held on the soil complex, by other basic cations. (See also CATION EXCHANGE CAPACITY.)

BENEFICIAL MICROBES—Microorganisms, such as *mycorrhizae* or *trichoderma*, which aid plant growth and development by either protecting the plant from disease organisms or increasing nutrient availability.

BENEFICIATION—A process that involves wet screening, hydroseparation, and concentration to remove impurities by flotation of the raw phosphate ore. The resulting product is dried and ground or slurried.

BEST MANAGEMENT PRACTICES (BMPs)—Those practices that combine scientific research with practical knowledge to optimize yields and increase crop quality while maintaining environmental integrity.

BIO-REMEDIATION—Use of living organisms to clean contaminated areas of target compounds.

BIURET—An undesireable by-product formed by the condensation of two molecules of urea during the urea manufacturing process.

BLENDED FERTILIZER—A physical mixture of different fertilizer materials.

BLOCKY—A soil structure characterized by approximately equal length in all three dimensions.

BLOOD MEAL—A high nitrogen analysis organic fertilizer source derived from the slaughter of cattle. Dried blood is completely soluble, rapidly mineralizes to plant-available forms, and is suitable for distribution through irrigation systems after it has been solubilized.

BONE MEAL—Cooked bones ground to a meal without any of the gelatin or glue removed. Steamed bone meal has been steamed under pressure to dissolve out part of the gelatin.

BRAND—The trade name assigned by a manufacturer to a particular fertilizer product.

BROADCAST APPLICATION—Uniformly distributing dry or liquid fertilizer over the entire desired area.

BROADCAST SPREADER—A machine for metering granular material from a feed hopper and distributing it over a wide area.

BUFFER CAPACITY OF SOIL—The ability of the soil to resist a change in pH (hydrogen ion concentration) when acid-forming or base-forming materials are added.

BULK BLEND—Physical mix of two or more dry fertilizer materials.

BULK BLENDING—The practice of mixing dry, individual, granular materials, or granulated bases. The resulting product is a mixture of granular materials rather than a granulated mixture.

BULK DENSITY—The weight of medium per volume.

CALCAREOUS SOIL—A soil containing calcium carbonate, or a soil alkaline in reaction because of the presence of calcium carbonate; a soil containing enough calcium carbonate to effervesce (fizz) when treated with dilute acid.

CALCINED CLAY—An inorganic component of soilless media mixes that is a granular substrate that resists compaction, does not float, and is moderately abrasive.

CALCITIC LIMESTONE—Calcium-based limestone.

CALCIUM CARBONATE EQUIVALENT (CCE)—The total neutralizing value of a material compared with pure calcium carbonate.

CALICHE—Cemented layers formed from illuviation of calcium carbonate.

CAPACITANCE SENSORS—Sensors that estimate soil moisture by measuring the apparent dielectric constant of the soil-water-air matrix.

CARBOHYDRATE—A group of neutral compounds of carbon, hydrogen, and oxygen (such as sugars, starches, and celluloses), most of which are synthesized by green plants. Carbohydrates constitute a major class of animal foods.

CARBON:NITROGEN RATIO—The ratio obtained by dividing the percentage of organic carbon by the percentage of nitrogen.

CATION—An ion carrying a positive charge of electricity. Common soil cations are calcium, magnesium, sodium, potassium, and hydrogen.

CATION EXCHANGE CAPACITY—The capacity of a medium to adsorb positively charged elements such as calcium (Ca^{2+}), magnesium (Mg^{2+}), ammonium (NH^{4+}), and potassium (K^{+}) onto the negatively-charged particle surfaces.

CELL—The basic structural and functional unit of an organism, a microscopic chemical factory that absorbs and secretes materials.

CHELATES—Certain organic chemicals, known as chelating agents, form ring compounds in which a polyvalent metal is held between two or more atoms. Such rings are chelates. Among the best chelating agents known are ethylenediaminetetraacetic acid (EDTA), hydroxyethylenediaminetriacetic acid (HEDTA), and diethylenetriaminepentaacetic acid (DTPA).

CHEMICAL AMENDMENT—Any substance used for the purpose of promoting plant growth or improving the quality of plants by conditioning soils solely through chemical means.

CHEMIGATION—Applying fertilizers and/or pesticides in irrigation waters to fertilize crops and control pests.

CHLOROPLASTS—Chlorophyll containing bodies that conduct photosynthesis.

CHLOROSIS—Yellowing of green portions of a plant, particularly the leaves.

CLAY—A minute soil particle less than 0.002 millimeter in diameter.

CLAYPAN—Cemented layers formed from clay particles.

CLEAR LIQUID SOLUTION—One or more plant nutrients in solution with no suspended particles.

COATED FERTILIZER—A term characterizing those products in which a water-soluble fertilizer granule is covered with a durable, permeable covering.

COIR—Natural fibers derived from coconut husks. It is the fibrous material found between the hard, internal shell and the outer coat of a coconut.

COLD WATER INSOLUBLE NITROGEN (CWIN)—That fraction of nitrogen contained in fertilizer not considered soluble in 68°F water. It is generally a measure of how much of the nitrogen is considered to show the slow release characteristic.

COLLOIDS—Soil particles (inorganic or organic) having small diameters ranging from 0.005 to 0.20 micron. Colloids are characterized by high ion exchange.

COLLUVIAL—Material transported by gravity.

COMPLETE FERTILIZER—A fertilizer containing all three of the primary fertilizer nutrients (nitrogen, phosphate, and potash) in sufficient amounts to be of value as nutrients.

COMPOST—Product of the facilitated decomposition of organic materials.

CONDITIONER (OF FERTILIZER)—A material added to a fertilizer to prevent caking and to keep it free-flowing.

CONSERVATION TILLAGE—Any system that leaves at least 30 percent of the soil surface covered with crop residue after planting.

CONTAINER CAPACITY—Water content of soilless media when a pot no longer freely drains following saturation.

CONTROLLED RELEASE FERTILIZER—"A fertilizer containing a plant nutrient in a form which delays its availability for plant uptake and use after application, or which extends its availability to the plant significantly longer than a reference 'rapidly available nutrient fertilizer' such as ammonium nitrate or urea, ammonium phosphate, or potassium chloride. Such delay of initial availability or extended time of continued availability may occur by a variety of mechanisms. These include controlled water solubility of the material (by semi-permeable coatings, occlusion, or by inherent water insolubility of polymers, natural nitrogenous organics, protein materials, other chemical forms), by slow hydrolysis of water soluble low molecular weight compounds, or by other unknown means." (AAPFCO)

CORTEX—In botany, central portion of a root or stem, which provides support and storage.

CRITICAL NUTRIENT RANGE (CNR)—That range of concentrations above which it is reasonably certain the crop is amply supplied with a selected nutrient and below which it is reasonably certain the crop is deficient. Also referred to as the *sufficiency range*.

CROTONYLIDENE DIUREA (CDU)—Produced from urea being acid catalyzed with crotonaldehyde or acetaldehyde.

CRYSTALLIZATION TEMPERATURE—Temperature at which a solution is saturated causing formation of solute crystals.

CURING—The process by which superphosphate or mixed fertilizers are stored until the chemical reactions have run to, or nearly to, completion.

CYTOPLASM—The portion of the protoplasm of a cell outside the nucleus.

DAMPING OFF—Sudden wilting and death of seedling plants resulting from attack by microorganisms.

DECISIEMENS PER METER (dS/m) or MILLIMHOS PER CENTIMETER (mmhos/cm)—Usually used to report electrical conductivity. Both are numerically equivalent units of measurement.

DEEP BANDING FERTILIZATION—Preplant application of nutrients placed 2 to 15 inches below the soil surface; applied in solid, fluid or gaseous forms.

DENITRIFICATION—The process by which anaerobic bacteria, usually in poorly aerated conditions, incorporate oxygen from nitrates into their bodies, resulting in the escape of nitrogen gas into air.

DISTRIBUTION PATTERN—The pattern of distribution of fertilizer on the soil from a mechanical applicator.

DISTRIBUTION UNIFORMITY—Variation or non-uniformity in the amount of irrigation water applied to locations within the irrigated area. Commonly expressed numerically as the average of the lowest one-fourth of the irrigation amounts divided by the average of the entire irrigation.

DOLOMITE—A material used for liming soils. Prepared by grinding dolomitic limestone, which contains both magnesium carbonate ($MgCO_3$) and calcium carbonate ($CaCO_3$) into fine particles. (See also LIME.)

DOLOMITIC LIMESTONE—Limestone that contains between 10 to 50 percent magnesium carbonate.

DRIBBLE FERTILIZATION—Dribbling or strip banding is a form of band placement that involves application of solid or fluid fertilizers in bands or strips of varying widths on the soil surface or on the surface of crop residues.

DROP SPREADER—A machine for metering granular material, from a feed hopper and distributing it directly below the spreader hopper only.

DUAL PLACEMENT—Placement of two fertilizer materials in subsurface bands.

ECOLOGY—The branch of biology that deals with the mutual relations among organisms and between organisms and their environment.

ELECTRICAL CONDUCTIVITY—A measure of the total soluble salts that are extractable from the medium.

ELECTRICAL RESISTANCE BLOCK—A sensor that estimates soil or media mositure content by measuring the electrical resistance across two wires embedded in a porous media in equilibrium with the surrounding media.

ELEMENTAL GUARANTEE—See GUARANTEE.

ENVIRONMENT—All external conditions that may act upon an organism or soil to influence its development, including sunlight, temperature, moisture, and other conditions.

ENZYMES—Protein substances produced by living cells that can change the rate of chemical reactions. They are organic catalysts.

EPIDERMIS—Outer layer of plant cells; generally protective in nature.

EROSION—The wearing away of the land surface by detachment and the transport of soil and rock materials through the action of moving water, wind, or other geological agents.

EVAPOTRANSPIRATION—The loss of water from a soil by evaporation and plant transpiration.

EXCHANGEABLE IONS—Ions held on the soil complex that may be replaced by other ions of like charge. Ions that are held so tightly that they cannot be exchanged are called *non-exchangeable*.

EXCHANGEABLE SODIUM PERCENTAGE—The degree of saturation of the soil exchange complex with sodium. It may be calculated by the formula:

$$ESP = \frac{\text{Exhangeable sodium (meq/100g soil)}}{\text{Cation exchange capacity (meq/100g soil)}} \times 100$$

EXPANDED SHALES, CLAYS, OR SLATES—Inorganic soilless media components made by super-heating flaky minerals in a rotary kiln to create lightweight, porous materials useful for improving aeration and water-holding capacity.

FALLOW—Land left idle in order to restore productivity, mainly through accumulation of water, nutrients, or both. Summer fallow is a common stage before planting in regions of limited rainfall. Extended fallow periods may help to control nematodes and diseases.

FEATHER MEAL—Organic nitrogen source consisting of finely ground poultry feathers originating as a by-product of the poultry industry.

FERTIGATION—Application of fertilizers through an irrigation system.

FERTILIZER—Any natural or manufactured material added to the soil to supply one or more plant nutrients. The term does not generally apply to soil amendments such as lime, gypsum, or elemental sulfur.

FERTILIZER FORMULA—The quantity and grade of materials used in making a fertilizer mixture.

FERTILIZER GRADE—An expression that indicates the weight percentage of plant nutrients in a fertilizer. Thus, a 10-20-10 grade contains 10 percent nitrogen (N), 20 percent phosphate (P_2O_5), and 10 percent potash (K_2O).

FERTILIZER PLACEMENT—Concentrating fertilizer into a band or strip at a specific location on or below the soil surface. Examples: starter, dribble fertilization, deep banding.

FERTILIZER RATIO—The relative proportions of primary nutrients in a fertilizer grade divided by the highest common divisor for that grade; e.g., grades 10-6-4 and 20-12-8 have the ratio 5-3-2.

FIBROUS ROOT—Root systems consisting of a group of numerous lone, slender roots of relatively equal width and length.

FIELD MOISTURE CAPACITY—The moisture content of soil in the field two or three days after a thorough wetting of the soil profile by rain or irrigation water. Field capacity is expressed as moisture percentage, dry-weight basis.

FISH EMULSION—A liquid organic fertilizer which is extracted from fish waste.

FISH MEAL—A dry organic fertilizer which is a processed waste by-product derived from fish wastes after fish emulsion has been extracted.

FIXATION—The process by which available plant nutrients are rendered unavailable or “fixed” in the soil. Generally, the process by which potassium, phosphorus, and ammonium are rendered unavailable in the soil. Also, the process by which free nitrogen is chemically combined, either naturally or synthetically. (See also REVERSION and NITROGEN FIXATION.)

FLOCCULATED SOIL—A soil in which structure is well aggregated; the condition is often achieved by adding amendments. A deflocculated soil would be indicative of a sodic soil in which poor structure is exhibited.

FLOCCULATION—The coagulation of individual soil particles due to the ions in solution. In most soils the clays and humic substances remain flocculated due to the presence of doubly and triply charged ions.

FLOWABILITY—The ability of the medium to move along conveyor belts or through other means of movement, such as a fertilizer spreader.

FLUID FERTILIZER—See LIQUID FERTILIZER and SUSPENSION FERTILIZER.

FOLIAR FERTILIZATION—Application of soluble fertilizers in the form of a spray to the foliage of plants.

FREE LIME—Native calcium carbonate.

FUNDAMENTAL TISSUES—Masses of cells that have little specialization in structure and function.

GENES—A section of DNA that codes for a specific physical trait.

GEOTROPISM—Growth response to gravity.

GRADE—The guaranteed analysis of a fertilizer containing one or more of the primary plant nutrient elements. Grades are stated in terms of the guaranteed percentages of nitrogen (N), phosphate (P_2O_5), and potash (K_2O), in that order. For example, a 10-10-10 grade would contain 10 percent nitrogen, 10 percent available phosphate, and 10 percent potash. (See also ANALYSIS.)

GRANULAR—A structure that includes all rounded aggregates.

GREENSAND—The name commonly applied to a sandy rock or sediment containing a high percentage of the green mineral glauconite and commonly used as an organic source of potassium.

GROUNDWATER—Water within the saturated zone of the earth that supplies wells and springs and is free to move under the influence of gravity.

GUANO—The decomposed dried excrement of birds and bats, used for fertilizer purposes. It is high in nitrogen and phosphate and at one time was a major fertilizer in this country.

GUARANTEE—The minimum percentage of plant nutrients claimed to be in a fertilizer.

GYPSUM ($CaSO_4 \bullet 2H_2O$)—The common name for calcium sulfate dihydrate, a mineral amendment used as a source of calcium and sulfur. Gypsum is used widely in reclaiming sodic soils in the western United States. Gypsum cannot be used as a liming material, but it may reduce the alkalinity of sodic soils by replacing sodium with calcium. When pure it contains approximately 18.6 percent sulfur and 23.3 percent calcium.

HARDNESS—A water measurement of divalent salts expressed as mg/L of calcium carbonate. An approximation would be (mg/L Ca $\times$ 2.5) + (mg/L Mg $\times$ 4.1).

HARDPAN—A hardened or cemented soil horizon or layer. The soil material may be sandy or clayey and may be cemented by iron oxide, silica, calcium carbonate, or other substances.

HEAVY METALS—Metallic elements in the transitional series of the periodic chart. They are not essential for plant nutrition and are usually found in small quantities in nature. They can be toxic to plants in high concentrations and to animals and humans if the concentration in the diet exceeds critical standards. Examples: cadmium, arsenic, lead, nickel, chromium and vanadium.

HEREDITY—The tendency for an offspring to display the characteristics of its parents.

HETEROTROPHIC BACTERIA—Single celled microorganisms that use organic carbon for growth.

HIDDEN HUNGER—A condition in a plant where there is a low level of plant nutrients but there are no observable nutrient deficiency symptoms.

HIGH PRESSURE INJECTION—A stream or pulse of fluid fertilizer forced below the surface at 2,000 to 6,000 pounds per square inch with prior opening of the soil by some mechanical means. Pressures are higher than in coulter injection.

HOMOGENOUS FERTILIZER—A fertilizer in which each granule or pellet has the same analysis.

HORIZON, SOIL—A layer of soil, approximately parallel to the soil surface, with distinct characteristics produced by soil-forming processes.

HUMIC ACID—The dark-colored organic material that can be extracted from the soil with dilute alkali and other reagents and that is precipitated by acidification to pH 1 to 2.

HUMUS—The well-decomposed, more-or-less stable portion of the organic matter in mineral soils. The stable decomposition residue from organic amendments.

HYDRAULIC CONDUCTIVITY—See PERMEABILITY, SOIL.

HYDROGEN ION CONCENTRATION—See pH.

HYDROLYSIS—Chemical decomposition in which a compound is broken down and changed into other compounds by the addition of water.

HYDROPHILIC—A dry medium which readily absorbs water, or "water loving".

HYDROPHOBIC—A dry medium which repels water.

HYDROPONICS—The growing of plants without the use of soil.

HYDROTROPISM—Growth response to water.

HYGROSCOPIC—A substance capable of taking up moisture from the air.

HYPNIUM—The deepest and most decomposed layer in a bog not usually used in the horticulture industry.

ILLUVIATION—The accumulation of many of the materials leached and transported from the surface soil that commonly accumulates in the subsoil (B horizon).

IMMOBILIZATION—The process of decomposing organic material with a high carbon nitrogen ratio that will temporarily tie up available nitrogen, reducing its availability to plants.

IMPREGNATION—Thorough mixing or spraying of a small amount of herbicide, fungicide, or other pesticide in a large amount of fertilizer.

INCORPORATION—Mechanical mixing of fertilizer with the surface soil; or mixing fertilizers into containers or potting mixes prior to planting or seeding.

INJECTION—Placement of fertilizer in the soil either through the use of pressure or nonpressure systems.

INORGANIC—Substances occurring as minerals in nature or obtainable from them by chemical means. Refers to all matter except the compounds of carbon, but includes carbonates.

INSOLUBLE—Not soluble. As applied to phosphate in fertilizer, that portion of the total phosphate that is soluble neither in water nor in neutral ammonium citrate. As applied to potash and nitrogen, not soluble in water.

INTEGRATED PEST MANAGEMENT—Judicious use of all available biological, physical, and chemical controls and crop rotations to reduce losses to crops caused by pests.

ION—An electrically charged particle. As used in soils, ion refers to an electrically charged element or combination of elements resulting from the breaking up of an electrolyte in solution. Since most soil solutions are very dilute, many of the salts exist as ions. For example, all or part of the potassium chloride (muriate of potash) in most soils exists as potassium ions and chloride ions. The positively charged potassium ion is a cation, and the negatively charged chloride ion is an anion.

IRRIGATION—The replacement or supplementation of rainfall with water from another source in order to grow plants.

IRRIGATION EFFICIENCY—Percentage of the total amount of irrigation water that is beneficially used. Beneficial uses include crop utilization, leaching for salinity control, and irrigation for climate (frost) control.

ISOBUTYLIDENE DIUREA (IBDU)—A combination of urea and isobutyraldehyde. This reaction forms a single oligomer, which is a white crystalline solid.

KELP—Any of several species of seaweed sometimes harvested for use as a fertilizer. Dried kelp will usually contain 1.6 to 3.3 percent N, 1 to 2 percent P_2O_5, and 15 to 20 percent K_2O.

KNIFED APPLICATION—Process where fertilizer materials are banded into the soil with a slender knifing tool.

LACUSTRINE—Soil that is uniform in texture but variable in chemical composition and that has been formed by deposits in lakes which have become extinct.

LEACHING—Also known as eluviation. The removal of materials in solution by the passage of water through soil.

LEACHING REQUIREMENT—Extra irrigation water required to push excess salinity beyond the root zone.

LEGUME—Plants including alfalfa, beans, lupins and clovers, which are known for their ability to fix atmospheric nitrogen into a form plants can use. This process is possible through the symbiotic relationship of rhizobacteria found in the root nodules of these plants.

LIME—Generally, the term *lime* or *agricultural* lime is applied to ground limestone (calcium carbonate), hydrated lime (calcium hydroxide), or burned lime (calcium oxide), with or without mixtures of magnesium carbonate, magnesium hydroxide, or magnesium oxide, and to materials such as basic slag, used as amendments to reduce the acidity of acid soils. In strict chemical terminology, *lime* refers to calcium oxide (CaO), but by an extension of meaning it is now used for all limestone-derived materials applied to neutralize acid soils.

LIME REQUIREMENT—The amount of standard ground limestone required to bring a 6.5-inch layer of an acre (about 2 million pounds in mineral soils) of acid soil to some specific lesser degree of acidity. In common practice, lime requirements are given in tons per acre of nearly pure limestone, ground finely enough so that all of it passes a 10-mesh screen and at least half of it passes a 100-mesh screen.

LIME SCORE—A measurement of lime quality. A function of the calcium carbonate equivalent (CCE), degree of fineness or particle size, and moisture.

LIQUID FERTILIZER—A fluid in which the plant nutrients are in true solution.

LOAM—The textural class name for soil having a moderate amount of sand, silt, and clay. Loam soils contain 7 to 27 percent clay, 28 to 50 percent silt, and less than 52 percent sand. (In the old literature, especially English literature, the term loam applied to mellow soils rich in organic matter, regardless of the texture. As used in the United States, the term refers only to the relative amounts of sand, silt, and clay; loam soils may or may not be mellow.)

LUXURY CONSUMPTION—The uptake by a plant of an essential nutrient in amounts exceeding what it needs. For example, if nitrogen is abundant in the soil, plants may take up more than is required and produce massive foliage.

MACRONUTRIENTS—Nutrients that plants require in relatively large amounts. Essential macronutrients are nitrogen, phosphorus, and potassium.

MANURE—The refuse from stables, barnyards and poultry or livestock operations, generally including both animal excreta and straw or other litter. In some other countries, the term *manure* is used more broadly and includes both farmyard or animal manure and "chemical manures," for which the term *fertilizer* is used in the United States.

MARL—An earthy deposit, consisting mainly of calcium carbonate, commonly mixed with clay or other impurities. It is formed chiefly at the margins of freshwater lakes. It is commonly used for liming acid soils.

MATRIC POTENTIAL—Adhesive forces due to the surface tension of water.

MAXIMUM ECONOMIC YIELD (MEY)—Yield at which unit costs of production are lowered to the point of highest net return per acre—the most profitable yield. The MEY is achieved through implementation of Best Management Practices.

MERISTEMATIC TISSUES—Cells that are composed of the embryonic undifferentiated cells capable of growth and division.

METHYLENE UREAS—A class of sparingly soluble fertilizers that contain predominantly intermediate chain-length polymers, primarily trimethylene tetraurea (TMTU) and tetramethylene pentaurea (TMPU).

MICROIRRIGATION—One of a number of closed irrigation systems characterized by low operating pressure (less than 40 psi), small orifice size, and construction, in part, from plastic materials. Examples are drip, microsprinkler, mister, bubbler, and fogger.

MICRONUTRIENTS—Nutrients that plants need in only small or trace amounts. Essential micronutrients are boron, chlorine, copper, iron, manganese, molybdenum, nickel, and zinc.

MILLIEQUIVALENT (meq)—A measurement of charge concentration. One-thousandth of an equivalent. In the case of sodium chloride, 1 meq would be 0.023 gram of sodium and 0.0355 gram of chloride in 1 liter of water. Milliequivalents per liter (meq/L) is the most useful method of reporting the major chemical components of water.

MILLIGRAMS PER LITER (mg/L)—Since one kilogram of water equals one liter, ppm is interchangeable with milligrams per liter (mg/L).

MINERALIZATION—Another term used to refer to the two processes, aminization and ammonification of organic matter to release available nitrogen.

MOSS—The topmost layer of a bog.

MOSS PEAT—Partially decomposed peat located directly under the moss layer in a bog that is very low in plant nutrients.

MUCK—Highly decomposed organic soil material developed from peat. Generally, muck has a higher mineral or ash content than peat and is decomposed to the point that the original plant parts cannot be identified.

MYCORRIZAL FUNGI—Soil fungal organisms that form a symbiotic association with plant root cells.

NECROSIS—Localized death of living tissue.

NEUTRON PROBE—Sensor that detects slowed neutrons that have reflected off hydrogen atoms in water molecules in the soil.

NITRIFICATION—The formation of nitrites and nitrates from ammonia (or ammonium compounds) in soils by microorganisms.

NITROGEN FIXATION—Generally, the conversion of free nitrogen to nitrogen compounds. Specifically in soils, the assimilation of free nitrogen from the soil air by soil organisms and the formation of nitrogen compounds that eventually become available to plants. The nitrogen-fixing organisms associated with legumes are called *symbiotic*; those not definitely associated with higher plants are *non-symbiotic* or *free-living*.

NONSALINE-SODIC SOIL—A soil that contains sufficient exchangeable sodium to interfere with the growth of most crop plants, but does not contain appreciable quantities of soluble salts. The exchangeable sodium percentage is greater than 15, the conductivity of the saturation extract is less than 4 decisiemens per meter (at 25°C), and the pH of the saturated soil paste usually ranges between 8.5 and 10.0.

NUTRIENT, ESSENTIAL—Any element taken in by a plant that is essential to its growth and is necessary for completion of its life cycle.

NUTRIENT RATIO—A ratio derived from plant or soil analyses when comparing the level of one element to another.

ORGAN—A group of tissues. In plants, organs are roots, stems, leaves, and reproductive organs.

ORGANIC—Relating or belonging to the class of chemical compounds having a carbon basis other than the inorganic carbonates.

ORGANICALLY CERTIFIED—A method of plant production that meets the requirements of an organic certification organization. Most organic plant production methods rely on non-synthetic inputs.

ORGANIC MATTER—Material that has come from a once-living organism; is capable of decay, or the product of decay; or is composed of organic compounds.

ORGANIC SOIL—A general term applied to soils or to soil horizons that consist primarily of organic matter, such as peat soils, muck soils, and peaty soil layers.

ORTHOPHOSPHATE—A salt of orthophosphoric acid, such as ammonium, calcium, or potassium phosphate. Each molecule contains a single atom of phosphorus.

ORTHOPHOSPHORIC ACID—H_3PO_4.

OSMOTIC POTENTIAL—The effect on the overall energy of water caused by the effect sugars, ions and other dissolved solutes have on water pressure.

PARENT MATERIAL—The unconsolidated mass of rock material (or peat) from which the soil profile develops.

PARTICLE DENSITY—The average density of the soil particles. Particle density is usually expressed in grams per cubic centimeter and is sometimes referred to as *real density* or *grain density*.

PARTS PER MILLION (ppm)—A notation for indicating small concentrations of materials. The expression gives the number of units by weight of the substance per million weight units of another substance. The term may be used to express the number of weight units of a substance per million weight units of a solution. The approximate weight of soil is 2 million pounds per acre–6 inches. Therefore, ppm × 2 equals pounds per acre–6 inches of soil, or ppm × 4 equals pounds per acre-foot of soil. Since one kilogram of water equals one liter, ppm is interchangeable with milligrams per liter (mg/L). The term "parts per million" is used to report constituents found in low concentrations in irrigation water such as iron, manganese, boron, nitrates, and nitrate-nitrogen.

PEAT—Plant residues that have accumulated and undergone varying amounts of decomposition in water or in excessively wet areas, such as swamps and bogs.

PERCOLATION—The downward movement of water through soil.

PERENNIAL—A plant that lives more than two years.

PERLITE—An inorganic media component that is a stable, white, bead-like particle that is produced from heat-treated volcanic rock.

PERMANENT WILTING POINT—The point when soil water is held so tightly by the soil or media that plants cannot extract it causing them to irreversibly wilt.

PERMEABILITY, SOIL—The quality of a soil horizon that enables water or air to move through it. It can be measured quantitatively in terms of rate of flow of water through a unit cross section in unit time under specified temperature and hydraulic conditions. Values for saturated soils usually are called *hydraulic conductivity*. The permeability of a soil is controlled by the least permeable horizon.

pH—A numerical designation of acidity and alkalinity. Technically, pH is the common logarithm of the reciprocal of the hydrogen ion concentration of a solution. A pH of 7.0 indicates precise neutrality; higher values indicate increasing alkalinity, and lower values indicate increasing acidity.

PHOSPHATE—A salt of phosphoric acid made by combining phosphoric acid with ions such as ammonium, calcium, potassium, or sodium.

PHOSPHATE ROCK—Phosphate-bearing ore composed largely of tricalcium phosphate. Phosphate rock can be treated with strong acids or heat to make available forms of phosphate. Finely ground rock phosphate is sometimes used in long-term fertility programs.

PHOTOPERIODISM—A plant's response to the hours of daylight or darkness.

PHOTOSYNTHESIS—The process by which green plants combine water and carbon dioxide to form carbohydrates under the action of light. Chlorophyll is required for the conversion of light energy into chemical energy.

PHOTOTROPISM—The response of plant growth to light.

PHYSICAL AMENDMENT—Any substance used for the purpose of promoting plant growth or improving the quality of plants by conditioning soils solely through physical means.

PLANT ANALYSIS CRITICAL LEVEL—A point where further increases in plant growth cease even though the tissue nutrient concentration is increasing.

PLANT AVAILABLE WATER—The amount of water stored between field or container capacity and permanent wilting point.

PLANT GROWTH—The increase in size, the enlargement, or the progressive development of a plant organism.

PLANT TISSUE ANALYSIS—Determination of chemical components of leaves or other plant parts to determine what nutrients are in the plant at the time of sampling.

PLATY—The name of a soil structure when the particles are arranged around a horizontal plane.

POLYPHOSPHATE—A salt of polyphosphoric acid, such as ammonium, calcium, or potassium polyphosphate. *Poly* means "many" and refers to multiple linkages of phosphorus in each molecule.

POLYPHOSPHORIC ACID—A phosphoric acid whose molecular structure contains more than one atom of phosphorus; generally a condensed phosphoric acid ranging in P_2O_5 content from 68 to 83 percent.

POP UP FERTILIZER—Application of low rates of fertilizer materials at planting in direct contact with the seed to encourage early rapid growth.

POROSITY—The fraction of soil volume not occupied by soil particles.

POSTPLANT FERTILIZER—Fertilizer applied after planting without specific references to method of application. Side-dressing, top-dressing and weed-and-feed applications are forms of postplant fertilization.

POTASH—AAPFCO has adopted the term *potash* to designate potassium oxide (K_2O).

POTENTIAL ACIDITY—A measure of hydrogen ions adsorbed to colloidal surfaces (CEC), also known as exchangeable acidity.

PRECISION FARMING—Carefully tailoring soil and crop management to fit the different conditions found in each level.

PREPLANT FERTILIZER—Fertilizer applied to soil prior to planting.

PRESSURE BOMBS—A device used to measure plant water potential by applying pressure to an excised leaf and observing the point at which sap appears at the petiole surface.

PRESSURE POTENTIAL—Also known as turgor, occurs when water enters a cell with a rigid wall causing the pressure to increase.

PRIMARY PLANT NUTRIENTS (PLANT FOODS)—Nitrogen (N), phosphate (P_2O_5), and potash (K_2O).

PRISM-LIKE—A soil structure where particles are arranged around a vertical line, bounded by relatively flat vertical surfaces.

PRODUCTIVITY—In simplest terms, the ability of the soil to produce. It differs from *fertility* to the extent that a soil may be fertile and yet unable to produce because of other limiting factors.

PROFILE, SOIL—A vertical section of the soil extending through all its horizons and into the parent material.

PROTECTIVE TISSUE—The epidermal or "skin" surface of a plant.

PROTEIN—Group of high-molecular-weight, nitrogen-containing compounds that yield amino acids on hydrolysis. Protein is a vital part of living matter and is one of the essential food substances of animals.

PROTOPLASM—The jellylike substance in plant and animal cells; it is basic to all life processes.

PSIG—Pounds per square inch, gauge reading.

PUDDLED SOIL—Dense, massive soil artificially compacted when wet and having no regular structure. The condition commonly results from the tillage of a clayey soil when it is wet.

PUMICE—An inorganic soilless media component that is ground, lightweight, porous volcanic rock with a sponge-like appearance.

PYRITE (FeS_2)—A mineral composed principally of iron and sulfur, with varying small amounts of other metals. "Fool's gold."

QUICK TESTS—Simple and rapid chemical tests of soils designed to give an approximation of the nutrients available to plants.

RATIO—See FERTILIZER RATIO.

RECLAMATION—The process of restoring lands to productivity by removing excess soluble salts or excess exchangeable sodium from soils.

REPRODUCTIVE ORGANS—Plant parts that include flowers and the resulting fruits.

RESPIRATION—The process where energy, stored in carbohydrates produced by photosynthesis, is converted and stored in the energy rich molecule ATP (adenosine triphosphate).

RETENTION ZONE—Soil zone where nutrients are concentrated following a fertilizer application. Usually refers to some sort of banded application.

REVERSION—The interaction of a plant nutrient with the soil that causes the nutrient to become less available. In fertilizer manufacturing, the excessive use of ammonia in ammoniation of phosphates results in phosphate reversion. (See also FIXATION.)

RHIZOSPHERE—Soil immediately surrounding the plant root.

RICE HULLS—A soilless media component that is the tough outer coatings of rice grains.

ROCK PHOSPHATE—An organic source of phosphorus, mainly poorly soluble apatite mineral.

ROCKWOOL—A substrate used in hydroponics made from melted basaltic rock that is spun into thread-type blocks.

ROOT—The plant organ that ordinarily grows downward into the soil, anchors the plant, and absorbs water, oxygen, and mineral nutrients.

SALINE-SODIC SOIL—A soil containing sufficient exchangeable sodium to interfere with the growth of most crop plants and containing appreciable quantities of soluble salts. The exchangeable sodium percentage is greater than 15, and the electrical conductivity of the saturation extract is greater than 4 decisiemens per meter (at 25°C). The pH reading of the saturated soil paste is usually less than 8.5.

SALINE SOIL—A soil containing enough soluble salts to impair its productivity for plants, but not containing an excess of exchangeable sodium.

SALT INDEX—A numerical comparison of fertilizer compounds using sodium nitrate's Salt Index 100 as the standard. Salt index compares relative osmotic potential at the plant surface from a solubilized fertilizer compound. The higher the salt index, the greater the potential for water movement out of the plant tissue, causing injury.

SALTING-OUT—The precipitation of dissolved salts when the temperature drops to the critical point.

SALTS—The products, other than water, of the reaction of an acid with a base. Salts commonly found in soils break up into cations (sodium, calcium, etc.) and anions (chloride, sulfate, etc.) when dissolved in water.

SAND—Individual rock or mineral fragments in soils having diameters ranging from 0.05 millimeter to 2.0 millimeters.

Usually sand grains consist chiefly of quartz, but they may be of any mineral composition. The textural class name of any soil that contains 85 percent or more sand and not more than 10 percent clay. It has low water- and nutrient-holding capacities. Used in media when additional weight is needed for container stability or to increase drainage.

SAP FLOW SENSORS—Devices that measure water flow in plant stems by determining the thermal dissipation of an amount of heat transferred to and carried by plant sap.

SATURATED SOIL PASTE—A particular mixture of soil and water commonly used for measurements and for obtaining soil extracts. At saturation, the soil paste glistens as it reflects light, flows slightly when the container is tipped, and slides freely and cleanly from a spatula for all soils, except those with high clay content.

SATURATION EXTRACT—The solution extracted from a soil at its saturation percentage.

SATURATION PERCENTAGE—The moisture percentage of a saturated soil paste, expressed on a dry-weight basis.

SEABIRD GUANO—An organic source of nitrogen derived from natural deposits of excrement and remains of birds living along extremely arid seacoasts.

SECONDARY PLANT NUTRIENTS—Calcium, magnesium, and sulfur.

SEDGE PEAT—Accumulates in swamps, along the edges of bogs, and beneath the sphagnum peat in bogs and has undergone longer decomposition than sphagnum peat.

SEGREGATION—Separation of one component or raw material from another, such as in a dry bulk blend.

SEPARATE, SOIL—One of the individual-size groups of mineral soil particles—sand, silt, or clay.

SERIES, SOIL—A group of soils that (1) have soil horizons similar in their differentiating characteristics and arrangement in the soil profile, except for the texture of the surface soil, and (2) are formed from a particular type of parent material. Soil series is an important category in detailed soil classification. Individual series are given proper names from places near the first recorded occurrence. Thus, names like Yolo, Panoche, Hanford, and San Joaquin are names of soil series that appear on soil maps, and each connotes a unique combination of many soil characteristics.

SEWAGE SLUDGE—An organic product resulting from the treatment of sewage. The composition varies widely depending on the method of treatment.

SHOOT—Another term for leaves or stems.

SIDE-BANDED FERTILIZER—Placement of fertilizer in bands on one or both sides of the seed or seedlings.

SIDE-DRESSED FERTILIZER—Application made to the side of crop rows after plant emergence.

SILT—(1) Individual mineral particles of soil that range in diameter between the upper size of clay, 0.002 mm, and the lower size of very fine sand, 0.05 mm. (2) Soil of the textural class silt containing 80 percent or more silt and less than 12 percent clay. (3) Water-deposited sediments in which the individual grains are approximately the size of silt, although the term is sometimes applied loosely to sediments containing considerable sand and clay.

SIMPLE FERTILIZER—Single-nutrient fertilizers.

SIZE GUIDE NUMBER—Particle size in millimeters at which 50 percent by weight of the sample is coarser and 50 percent is finer, times 100.

SLOW RELEASE—See CONTROLLED RELEASE.

SLUDGES—The residues from sewage treatment facilities.

SLURRY FERTILIZER—A fluid mixture containing dissolved and undissolved plant nutrient materials that requires continuous mechanical agitation to assure homogeneity.

SODIC SOIL—A soil that contains sufficient exchangeable sodium to interfere with the growth of most plants, either with or without appreciable quantities of soluble salts. (See also NONSALINE-SODIC SOIL and SALINE-SODIC SOIL.)

SODIUM ADSORPTION RATIO—A ratio for soil extracts and irrigation waters used to express the relative activity of sodium ions in exchange reactions with soil. The ionic concentrations are expressed in milliequivalents per liter.

SODIUM PERCENTAGE—The percent sodium of total cations. Calculations are based on milliequivalents rather than weight.

SOIL—A complex natural material derived from disintegrated, decomposed, and reformed minerals and organic matter that provides nutrients, moisture, and anchorage for land plants.

SOIL INJECTION—The application of fertilizers below the surface of the soil.

SOILLESS MEDIA—Media used for growing plants in containers; made of a combination of lightweight, porous substrates and little, if any, soil.

SOIL MOISTURE STRESS—The sum of the soil moisture tension and the osmotic pressure of the soil solution. It is the force plants must overcome to withdraw moisture from the soil.

SOIL MOISTURE TENSION—The force by which moisture is held in the soil. It is a negative pressure and may be expressed in any convenient pressure unit. Tension does not include osmotic pressure values.

SOIL SEPARATES—Three categories of individual mineral particles: sand, silt, and clay.

SOIL STRUCTURE—The way in which soil particles are grouped together into larger shapes.

SOIL TEST CRITICAL LEVEL—The point where increases in plant growth cease in response to further increases in the concentration of a particular soil nutrient.

SPECIMEN TREES OR SHRUBS—Plants grown by themselves in a garden or lawn that are the focal point of a landscape.

SPHAGNUM PEAT—A layer that is partially decomposed and lies below moss peat in a bog.

SPLIT APPLICATION—Fertilizer applied two or more times during the crop growing season; pre-plant and one or more postplant applications are common.

STABILIZED NITROGEN—The use of chemical additives or inhibitors to slow the transformation of nitrogen resulting in an extended period of availability.

STARTER FERTILIZER—Fertilizer applied at planting either in direct seed contact or to the side and below the seed. Exact position is not implied.

STELE—Plant structure that provides a means of transport for water and nutrients to and from the roots.

STOMATE—A pore on the surface of a plant that regulates the outflow of water vapor and oxygen and the inflow of carbon dioxide.

STRUCTURE, SOIL—The physical arrangement of the soil particles.

SUBSOIL—Roughly, that part of the soil below plow depth.

SUCTION—The energy required to remove water from the surface of a particle.

SUPERPHOSPHATE—Superphosphate is a product obtained by mixing rock phosphate with either sulfuric acid or phosphoric acid or with both acids. The grade that shows the available phosphate shall be used as a prefix to the name. Example: 20 percent superphosphate. (AAPFCO)

SUPERPHOSPHORIC ACID—See POLYPHOSPHORIC ACID.

SURFACE BAND APPLICATION—Placement of a liquid or solid fertilizer as either a dribble or forced stream on the soil surface.

SURFACE WATER—Water on or above the ground, including rivers, lakes, canals, and reservoirs.

SUSPENSION FERTILIZER—A fluid containing dissolved and undissolved plant nutrients. The suspension of the undissolved plant nutrients may be inherent to the materials or may be produced with the aid of a suspending agent of non-fertilizer properties. Mechanical agitation may be necessary in some cases to facilitate uniform suspension of undissolved plant nutrients.

SUSTAINABLE AGRICULTURE—The production of crops with judicious use of all inputs to maintain production indefinitely.

SYMBIOSIS—The living together of two different organisms with a resulting mutual benefit. A common example is the association of rhizobia with legumes; the resulting nitrogen fixation is sometimes called *symbiotic nitrogen fixation*. Adjective: symbiotic.

TANKAGE—Dried animal residue. Process tankage is made from leather scrap, wool, and other inert nitrogenous materials by steaming under pressure with or without addition of acid. This treatment increases the availability of the nitrogen to plants.

TAPROOT—Root systems consisting of one large downward root with several lateral roots.

TENSIOMETER—A device used to measure the tension with which water is held in the soil; directly measures soil suction.

TEXTURE, SOIL—The relative proportions of the various size groups of individual soil grains in a mass of soil. Specifically, the proportions of sand, silt, and clay.

THATCH—The remains of dead and living turf grass stems that build up between the green blades and the soil surface. Excessive thatch can lead to poor water infiltration and aeration leading to disease and insect problems in turf.

THERMAL DISSIPATION SENSORS—A mositure sensor that indirectly measures soil water content by measuring the dissipation of heat from a porous ceramic block or disk in contact with soil or media.

THERMOTROPISM—Plant growth response to temperature.

TILTH—The physical condition of a soil with respect to its fitness for the growth of plants.

TISSUE—A group of cells that function as a unit. In plants, tissues are classified as meristematic, fundamental, protective, and vascular.

TOPDRESSING—An application of an amendment or fertilizer over an area and not physically incorporated. In turfgrass, topdressing refers to the application of a sand, organic, or other material in light applications to the turf. Nursery topdressing usually refers to a fertilizer application that is applied to the surface inside a container and not incorporated.

TOPOGRAPHY—The shape and position of land surfaces.

TOTAL ACIDITY—The combination of potential and active acidity.

TOTAL DISSOLVED SOLIDS—A measure of the total salt content of water measured by letting a known weight of water evaporate completely and weighing the salts that remain. It is usually reported as parts per million (ppm).

TRACE ELEMENTS—See MICRONUTRIENTS.

TRANSPIRATION—Loss of water vapor from the leaves and stems of living plants to the atmosphere.

TURFGRASS—A plot of grass, usually mowed, as one around a residence or in a park or sports field.

UNIFORMITY INDEX—The particle size at which 95 percent of the material is retained, divided by the particle size at which 10 percent of the material is retained, multiplied by 100.

UREAFORM—A slow release nitrogen that is a polymer of urea and formaldehyde that contains a minimum of 35 percent total nitrogen, with at least 60 percent of it as Cold Water Insoluble Nitrogen (CWIN).

VAPOR PRESSURE—The pressure exerted above a liquid because of the tendency of vapor to escape from the surface. Typical examples are the pressure above liquid anhydrous ammonia or ammonia-ammonium nitrate solutions. A negative value indicates that the vapor pressure above the liquid is less than atmospheric pressure. Vapor pressure is temperature dependent. Increasing the temperature increases the vapor pressure above the liquid.

VARIABLE-RATE TECHNOLOGY—Application equipment designed to apply different rates of crop inputs depending on geographic location in a field.

VASCULAR TISSUE—Function in the conductive processes of the plant; xylem and phloem.

VOLATILIZATION—The evaporation or changing of a substance from liquid to vapor.

WATER TABLE—The upper surface of groundwater.

WATER TABLE, PERCHED—The upper surface of a body of free groundwater in a zone of saturation separated from underlying groundwater by unsaturated material.

WEATHERING—The physical and chemical disintegration and decomposition of parent materials as in soil formation.

WEED-AND-FEED—A term used for a product that is a combination of fertilizer and herbicide that can be applied in one application.

WET-PROCESS ORTHOPHOSPHORIC ACID—A liquid phosphate fertilizer produced by reacting sulfuric acid with finely ground phosphate rock.

Index

Note: Page numbers with *f* indicate figures; those with *t* indicate tables.

D

I

J

K

L

N

T